M000121808

Wildland Water Quality Sampling and Analysis

WILDLAND WATER QUALITY SAMPLING AND ANALYSIS

John D. Stednick

Department of Earth Resources
College of Forestry and Natural Resources
Colorado State University
Fort Collins, Colorado

ACADEMIC PRESS, INC.
Harcourt Brace Jovanovich, Publishers
San Diego   New York   Boston
London   Sydney   Tokyo   Toronto

Cover photograph by John D. Stednick.

This book is printed on acid-free paper. ∞

Copyright © 1991 by Academic Press, Inc.
All Rights Reserved.
No part of this publication may be reproduced or transmitted in any
form or by any means, electronic or mechanical, including photo-
copy, recording, or any information storage and retrieval system,
without permission in writing from the publisher.

Academic Press, Inc.
San Diego, California 92101

United Kingdom Edition published by
Academic Press Limited
24–28 Oval Road, London NW1 7DX

Library of Congress Cataloging-in-Publication Data

Stednick, John D.
    Wildland water quality sampling and analysis / John D.
Stednick.
        p.   cm.
    ISBN   0-12-664100-5   (alk. paper)
    1. Water quality--Measurement--Laboratory manuals. 2. Water
chemistry--Laboratory manuals.   I. Title.
    TD367.S75   1991
    628.1'61--dc20   90-39858
                                                               CIP

Printed in the United States of America
90 91 92 93   9 8 7 6 5 4 3 2 1

Contents

Chapter 18

Enteric Bacteria   183

Chapter 19

Contract Laboratory Selection   195

Chapter 20

Data Analysis and Presentation   199

Index   215

Preface

Any land-use activity has the potential to affect that basin's water quality. Previous land-use activities were often for the extraction or production of a single natural resource. Watershed management was often a corrective rather than preventative measure. Recognition of potential non-point-source pollution from land-use activities allows watershed management to become a positive facet of land-use activity planning.

The passage of the Federal Water Pollution Control Act Amendments in 1972 directed state assessments of non-point-source pollution from land-use activities and development of watershed management practices to minimize this pollution. These best management practices (BMPs) are meant to meet state water-quality standards. The Clean Water Act of 1987 directs states to further define non-point-source pollution and to identify cumulative effects from multiple (in time and space) land-use activities.

With continued land-use activity, it is essential to recognize potential, and quantify actual, changes in the quality of our finite water resources. Proper water-quality sampling techniques and chemical analyses are necessary to obtain valid results. These results, when documented through appropriate data presentation, statistics, interpretation, and discussion allow for continued refinement of BMPs and add to the existing water-quality database.

Water-quality data are only as good as the water quality sampling. Specific sections in this book address water-quality sampling in precipitation, surface waters, and groundwaters. Analytical techniques for common water-quality variables, detailed by chapter, include typical values for wildland waters. The model laboratory exercise sheets help identify analytical and system variability, from the calculated result to the reported value.

This book can serve as a text for upper division undergraduate and graduate student courses in water quality and as a reference for the professional in wildland water quality. The methodologies and techniques discussed are for non-point-source water quality, but are applicable to point-source monitoring.

This book is the result of a series of class handouts developed over time, for a course in land-use and water quality at Colorado State University. Several of my graduate students, now colleagues, made contributions to this effort. They are Lisa Roig, Don Campbell, Steve Johnson, Mike DeWeese, Julianne Thompson, Jay Schug, and Bill Weiss. Barbara Holtz did the word processing and the updates and revisions. And thanks to Susan.

John D. Stednick

Chapter 1

Chemistry Review

General Chemical Relationships

ATOMIC AND MOLECULAR WEIGHTS

The atomic weight of an element is given in atomic mass units (AMU). The molecular weight of a compound is the sum of the weight of its component atoms. An example of a molecular weight computation is given for H_2SO_4 (sulfuric acid):

Atomic Mass Units

$2 \times$ Hydrogen	$2 \times 1.008 =$	2.016 AMU
$1 \times$ Sulfur	$1 \times 32.060 =$	32.060 AMU
$4 \times$ Oxygen	$4 \times 16.000 =$	64.000 AMU

Molecular weight = 98.076 AMU

The molecular weight of a compound is also numerically equal to the number of grams per mole of pure compound. There are 98.076 grams of H_2SO_4 per mole of H_2SO_4. Remember that there are 6.022169×10^{23} molecules per mole of compound (Avogadro's number). Atomic weights of all the elements are shown in Figure 1.1.

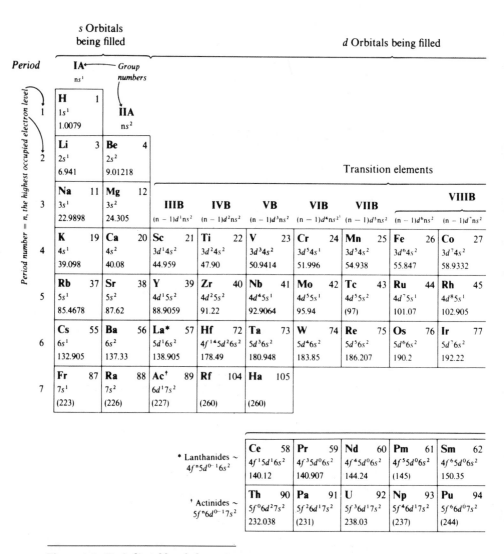

Figure 1.1 Periodic table of elements.

p Orbitals being filled

$(n-1)d^8ns^2$	IB $(n-1)d^{10}ns^1$	IIB $(n-1)d^{10}ns^2$	IIIA ns^2np^1	IVA ns^2np^2	VA ns^2np^3	VIA ns^2np^4	VIIA ns^2np^5	Noble gases VIIIA ns^2np^6
								He 2 $1s^2$ 4.0026
			B 5 $2s^22p^1$ 10.81	**C** 6 $2s^22p^2$ 12.011	**N** 7 $2s^22p^3$ 14.0067	**O** 8 $2s^22p^4$ 15.9994	**F** 9 $2s^22p^5$ 18.9984	**Ne** 10 $2s^22p^6$ 20.179
			Al 13 $3s^23p^1$ 26.9815	**Si** 14 $3s^23p^2$ 28.086	**P** 15 $3s^23p^3$ 30.9738	**S** 16 $3s^23p^4$ 32.06	**Cl** 17 $3s^23p^5$ 35.453	**Ar** 18 $3s^23p^6$ 39.948
Ni 28 $3d^84s^2$ 58.70	**Cu** 29 $3d^{10}4s^1$ 63.546	**Zn** 30 $3d^{10}4s^2$ 65.38	**Ga** 31 $4s^24p^1$ 69.72	**Ge** 32 $4s^24p^2$ 72.59	**As** 33 $4s^24p^3$ 74.9216	**Se** 34 $4s^24p^4$ 78.96	**Br** 35 $4s^24p^5$ 79.904	**Kr** 36 $4s^24p^6$ 83.80
Pd 46 $4d^{10}$ 106.4	**Ag** 47 $4d^{10}5s^1$ 107.868	**Cd** 48 $4d^{10}5s^2$ 112.40	**In** 49 $5s^25p^1$ 114.82	**Sn** 50 $5s^25p^2$ 118.69	**Sb** 51 $5s^25p^3$ 121.75	**Te** 52 $5s^25p^4$ 127.60	**I** 53 $5s^25p^5$ 126.904	**Xe** 54 $5s^25p^6$ 131.30
Pt 78 $5d^96s^1$ 195.09	**Au** 79 $5d^{10}6s^1$ 196.967	**Hg** 80 $5d^{10}6s^2$ 200.59	**Tl** 81 $6s^26p^1$ 204.37	**Pb** 82 $6s^26p^2$ 207.19	**Bi** 83 $6s^26p^3$ 208.980	**Po** 84 $6s^26p^4$ (209)	**At** 85 $6s^26p^5$ (210)	**Rn** 86 $6s^26p^6$ (222)

f Orbitals being filled

Eu 63 $4f^75d^06s^2$ 151.96	**Gd** 64 $4f^75d^16s^2$ 157.25	**Tb** 65 $4f^95d^06s^2$ 158.925	**Dy** 66 $4f^{10}5d^06s^2$ 162.50	**Ho** 67 $4f^{11}5d^06s^2$ 164.930	**Er** 68 $4f^{12}5d^06s^2$ 167.26	**Tm** 69 $4f^{13}5d^06s^2$ 168.934	**Yb** 70 $4f^{14}5d^06s^2$ 173.04	**Lu** 71 $4f^{14}5d^16s^2$ 174.97
Am 95 $5f^76d^07s^2$ (243)	**Cm** 96 $5f^76d^17s^2$ (247)	**Bk** 97 $5f^96d^07s^2$ (247)	**Cf** 98 $5f^{10}6d^07s^2$ (251)	**Es** 99 $5f^{11}6d^07s^2$ (254)	**Fm** 100 $5f^{12}6d^07s^2$ (257)	**Md** 101 $5f^{13}6d^07s^2$ (258)	**No** 102 $5f^{14}6d^07s^2$ (255)	**Lr** 103 $5f^{14}6d^17s^2$ (260)

SOLUTIONS

Some of the more important determinations in water chemistry are dependent on the use of standard solutions. A standard solution contains a known weight of the active substance dissolved in a definite volume of solution. Methods involving the use of such solutions are known as "volumetric procedures," since the quantitative result is obtained by measurement of volumes.

It is convenient to have solutions of known concentrations for use in various determinations. Such solutions simplify the calculation of results, a decided advantage in routine analysis. They can best be prepared by diluting portions of standard stock solutions in such a manner as to give solutions of the desired strength.

The molarity of a solute is the number of moles (mol) of solute per liter of solution and is usually designated by M. A 6.0-molar solution of H_2SO_4 is labeled 6.0 M. The label means that the solution has been made up in a ratio that corresponds to adding 6.0 mol of H_2SO_4 (588.456 g) to enough water to make a liter of solution.

Molar concentration may be defined as the number of moles of solute per liter of solution:

$$M = \frac{\text{number of moles of solute}}{\text{number of liters of solution}}$$

Thus, a 1 molar solution (M) contains 1 mol of solute per liter of solution, and 1 L (liter) of a 1 M solution contains 1 mole of solute. Also, 1 L of a 2 M solution contains 2 mol of solute, 1 L of a 0.5 M solution contains 0.5 mol of solute, 2 L of 1.5 M solution contains 3 mol of solute, and 0.5 L of a 0.2 M solution contains 0.1 mol of solute, or

(Number of liters of solution) × (molarity of solution)
= number of moles of solute

or

Number of liters solution × M = number of moles of solute

The molality of a solute is the number of moles of solute per kilogram of solvent. It is usually designated by a lowercase m. The label 6.0 m H_2SO_4 is read "6.0 molal" and represents a solution made by adding, to every 6.0 moles of H_2SO_4, 1000 g of water. Molality is rarely used today.

The strength of a standard solution is usually expressed in terms of its normality. A one normal (1 N) solution contains one equivalent weight of the active substance in one liter of solution. The equivalent weight of a substance is the weight of that substance that will react with 1 gram of hydrogen.

$\frac{6 \text{ eq}}{1 \text{ liter}}$

EQUIVALENT WEIGHTS

Chemical reactions are based on equivalent (electron) exchanges regardless of molecular weight. There are two general types of chemical reactions encountered in the volumetric determinations used for water analysis: (1) simple neutralization or double-decomposition reactions, and (2) reactions involving oxidation and reduction.

The reactions of the first type involve a change in position of the various atoms and groups making up the reacting substances. The following equation illustrates this type of reaction.

$$HCl + NaOH \leftrightharpoons NaCl + H_2O$$

In a reaction of this type, the equivalent weight of each compound reacting is calculated by dividing the molecular weight of the compound by the number of replaceable hydrogen atoms or their equivalent in that compound.

$$\frac{\text{Molecular weight of a substance}}{\text{Number of replaceable hydrogen atoms}} = \frac{\text{equivalent weight of the}}{\text{substance}}$$

Thus the equivalent weight of H_2SO_4 is 49.04, since the molecular weight (98.076) is divided by 2, because there are two replaceable hydrogen atoms. Thus a 1 M solution of H_2SO_4 is the same as a 2 N solution of H_2SO_4. In NaOH (40.0 AMU), one sodium atom can be replaced by one hydrogen atom and is equivalent to one atom of hydrogen. Therefore, the equivalent weight of NaOH is 40.0.

The second type of reaction, oxidation–reduction, is illustrated by the following equation:

$$2KMnO_4 + 5H_2C_2O_4 + 3H_2SO_4 \leftrightharpoons 2MnSO_4 + K_2SO_4 + 10CO_2 + 8H_2O$$

A study of this equation will show that there is more involved than a simple rearrangement of atoms and groups. Manganese, for instance, has a positive valence of seven in $KMnO_4$, but only two in $MnSO_4$. The Mn has lost five positive valences (has been reduced);

$1 M = 1 N$ for HCl

$6 N$

since there are two Mn atoms, the total loss is 10 valences. Likewise, each carbon atom in $H_2C_2O_4$ has a valence of positive three (3+), while each C atom in CO_2 has valence of positive four (4+). The 10 C atoms thus gain a total of 10 valences (one for each C atom). The total loss of positive valence by one kind of atom in this type of reaction must equal the total gain of positive valences by another kind of atom.

The equivalent weight of a compound entering into an oxidation–reduction reaction is equal to its molecular weight divided by the valence of the element in that compound.

$$\frac{\text{Molecular weight}}{\text{Valence}} = \text{equivalent weight}$$

$$SO_4^{2-} = 96.060 \text{ AMU}$$

$$\text{Valence} = 2$$

$$\frac{96.060}{2} = 48.030 \text{ equivalent weight}$$

$$\frac{\text{Concentration}}{\text{Equivalent weight}} = \text{equivalent concentration}$$

$$\frac{10 \text{ mg L}^{-1} SO_4^{2-}}{48.030} = 0.210 \text{ meq L}^{-1} SO_4^{2-}$$

Common expressions of concentrations include milliequivalents per liter (meq L^{-1}).

The normality of a standard solution is the ratio of the weight in grams of the substance in one liter to the equivalent weight.

$$\frac{\text{Weight in grams per liter}}{\text{Equivalent weight}} = \text{normality (N)}$$

UNITS OF EXPRESSION

The compounds of nitrogen, phosphorus, and sulfur may be expressed as a concentration of the compound or as the elemental form. For example, NO_3–N nitrate as nitrogen as opposed to NO_3^-

nitrate. The conversion of concentrations is the weight of the compound to the weight of the elemental form. To convert between expressions:

$$NO_3^- = \text{assume } 45 \text{ mg L}^{-1}$$

$$NO_3-N = ?$$

$$NO_3-N = 62 \text{ AMU}$$

$$N = 14 \text{ AMU}$$

$$\frac{14}{62} \times 45 \text{ mg L}^{-1} = 10 \text{ mg L}^{-1} \text{ as } NO_3 - N$$

Research in water chemistry often involves analysis for constituents with concentrations below milligrams per liter. The use of $\mu g \text{ L}^{-1}$ (micrograms per liter) may create problems with data presentations or abuse of significant figures. A commonly used alternative is to express the concentrations as $\mu eq \text{ L}^{-1}$ (microequivalents per liter) or as mmol L^{-1} (millimoles per liter).

Review of Solution Chemistry

Much of the geochemistry of the earth's surface is involved with solutions (streams, lakes, groundwater, ocean, etc.). The temperatures and pressures at which reactions occur in these environments are approximately equal to those occurring in the chemical laboratory [25°C, 1 atmosphere (atm) of pressure]. It is therefore appropriate to review some of the principles developed by solution chemists, with the realization that these principles apply directly to geochemical processes taking place at the surface of the earth or at shallow depths in aqueous natural environments.

COLLIGATIVE PROPERTIES OF SOLUTIONS

Such properties as the lowering of the freezing temperature or the raising of the boiling point of a solvent by a solute are termed "colligative properties" of the solution. They are considered to be simple due to the dilution of the solvent by the solute. An example is the vapor pressure of alcohol. As water is added to alcohol, fewer molecules of alcohol are available at the surface of the solution, and consequently the pressure exerted by alcohol molecules escaping from the surface is reduced. The drop in alcohol vapor pressure can be considered to be due to dilution of the alcohol with water.

The various colligative properties are all related, and one can be derived from the other by simple mathematics. The colligative properties are also independent of the particular solute (if the solutes being considered are nonelectrolytes). In other words, it wouldn't matter whether one mole of solute A or one mole of solute B were added to the solution—both would lower the freezing point by the same amount, because of Avogadro's constant number of molecules per mole.

ELECTROLYTES

The term electrolyte was originally given to substances that, when dissolved, yielded a conducting solution—indicating that charged particles were present (an ionic solution).

The measurement of the properties of electrolytes gave what at first seemed surprising results. For NaCl, for example, the predicted freezing point depression of a 0.1-molal solution is 0.186°C. A factor termed the "van't Hoff factor" (i) is defined as the ratio of observed to expected values. For the NaCl solution above, i is 1.86. For 0.01 and 0.001 molal solution of NaCl, i is, respectively, 1.94 and 1.97. The value for i can be obtained from measurement of any of the colligative properties, and is the same for each. The i is close to the number of ions that can be derived from one "molecule" by dissociation. Common table salt, NaCl, dissociates into Na^+ and Cl^-.

When a nonelectrolyte is put into solution, each molecule acts as a particle that dilutes the solution and produces colligative effects such as lowering of the freezing temperature. When an electrolyte such as NaCl is added, each molecule of NaCl splits into two ions, each of which acts as a diluting particle, and the colligative property is roughly doubled.

The fact that values for i only approached whole numbers but were not actually whole numbers was taken to mean that there was almost, but not quite, complete dissociation. Some compounds were found to have i values that were even farther removed from whole numbers and did not approach a whole number as dilution was increased. These were termed "weak electrolytes" (acetic acid, ammonium hydroxide), in contrast to salts such as NaCl, which were called "strong electrolytes." The idea evolved that there was an equilibrium between a molecule of the compound and its dissociation products. According to this theory, there were actually molecules of NaCl in equilibrium with Na^+ and Cl^- ions of the solution.

This theory is incorrect, at least for strong electrolytes. Modern studies of solutions have developed the concept of charged ions attracting rather large clusters of water molecules. A steady-state equilibrium as pictured in the old theory would involve a continuous stripping and reassembling of the clusters. This seems rather unlikely. (It should be pointed out that some solutes do remain in solution as discrete molecules—for example HCl, which is covalently bonded.)

ACTIVITY COEFFICIENTS

This theory assumes that the electrolyte is completely dissociated. Because of the charge on an ion, the probability is greater that it will be in the neighborhood of an ion of opposite charge rather than one of like charge. Each charged ion water cluster will be surrounded by an "ionic atmosphere" of opposite sign ions that will immobilize the ion more tightly than if the "atmosphere" were absent. The result is a restriction of the freedom of motion of the ion. The greater the concentration of the solution, the greater will be the immobilization of all ions.

Because of these interactions, the behavior of the electrolyte is different from that predicted from its actual concentration. It is desirable to define an "effective concentration."

An effective concentration or activity is calculated by the Debye–Hückel equation for the "activity coefficient," which, when multiplied by the concentration, yields the effective or "thermodynamic concentration." In other words, because of the restrictions on the freedom of the ion, the electrolyte behaves as if there were less of it present than there actually is in solution. The activity coefficient simply tells us what fraction of the actual concentration should be used as the thermodynamic concentration.

The Debye–Hückel equation is

$$\log \gamma_i = \frac{-AZ_i^2\sqrt{I}}{1 + Ba_0\sqrt{I}}$$

which at 25°C temperature and 1 atm pressure $A = 0.5085$ and $B = 0.3281$.

$$\log\gamma_i = -0.51 z_i^2 \sqrt{I}$$

where γ_i = activity coefficient of the ion in solution; A, B = constants related to temperature and pressure; a_0 = variable related to hydrated ratio of the ion; Z_i = charge of the ion; and I = ionic strength.

The ionic strength of a solution is a measure of the strength of the electrostatic field caused by the ions and is computed as follows:

$$I = \tfrac{1}{2}\Sigma m_i \, Z_i^2$$

The molar concentration of each ionic species is multiplied by the square of the charge (Z). (Each ion is therefore assigned a weight that is proportional to the square of its charge.) The result, both theoretical and experimental, is that ions of multiple charge have a greater effect on activity coefficients than ions with single charges.

Examples: What are the activities of Ca^{2+} and Na^+ in a solution that is 0.001 molar in Na_2SO_4 and 0.005 molar in $CaCl_2$?

$$I = \tfrac{1}{2}\Sigma m_i \, Z_i^2$$

The ionic strength is

Na^+	$0.002\ M \times 1^2$	0.002
Ca^{2+}	$0.005\ M \times 2^2$	0.020
SO_4^{2-}	$0.001\ M \times 2^2$	0.004
Cl^-	$0.010\ M \times 1^2$	0.010
		0.036

$$\frac{0.036}{2} = 0.018$$

$$I = 0.018$$

$$\log \gamma\, Ca^{2+} = -0.51 \times 2^2 (0.018)^{1/2} = -0.274$$

$$\gamma\, Ca^{2+} = 0.532$$

$$[Ca^{2+}] = (\gamma\, Ca^{2+})(Ca^{2+}) = (0.532)(0.005) = 0.00626$$

.00266

$$\log \gamma \, Na^+ = -0.51 \times 1^2 \, (0.018)^{1/2} = -0.068$$

$$\gamma \, Na^+ = 0.854$$

$$[Na^+] = (\gamma \, Na^+)(Na^+) = (0.854)(0.002) = 0.00171$$

Activity are denoted by [ ]; concentrations by ( ).

These activity values can be used in solubility calculations. The upper limit of usefulness of activity values is an ionic strength of 0.1 when only univalent ions are involved, less when multivalent ions are involved.

SOLUBILITY PRODUCT

When a solution is saturated, solid material in excess remains in a separate phase. There is an equilibrium between the excess solid and the ions in solution:

$$AB(s) \leftrightharpoons A^+(aq) + B^-(aq)$$

An equilibrium constant expression can be written for molar activities as products over reactants:

$$K_{sp} = \frac{[A^+][B^-]}{[AB]}$$

The solid has a definite density and therefore a definite (constant) concentration. The "solubility product," K_{sp}, usually is defined at 25°C and 1 atm.

For electrolytes that produce more than one ion of a given charge, the concentration is raised to the power corresponding to the number of ions of that charge:

$$Ca_3 \, (PO_4)_2 \leftrightharpoons 3Ca^{2+} + 2PO_4^{3-}$$

$$K_{sp} = [Ca^{2+}]^3[PO_4^{3+}]^2$$

The solubility product equation is useful in that it describes the saturation condition at standard conditions not only when just one solute is present, but when several solutes are present and having an ion in common. Consider a solution initially saturated with lead sulfate. The solubility product of $PbSO_4$ is 1×10^{-8}; therefore, the molar concentrations of lead ions and sulfate ions, which are equal, are each 1×10^{-4}, since the dissociation is 1 mol of Pb^{2+} to 1 mol of SO_4^{2-}. Now make the solution 0.1 molar in sodium sulfate.

$$K_{sp} = 1 \times 10^{-8} = [Pb^{2+}] \; (10^{-1})$$

$$[Pb^{2+}] = 10^{-7}$$

Originally, $[Pb^{2+}]$ was 10^{-4}. Therefore, most of the lead was precipitated.

Solubility products allow one to predict the order of precipitation. Suppose that sodium chromate (Na_2CrO_2) is being added slowly to a solution that is 0.1 molar in Ag^+ and 0.02 M in Ba^{2+}. The solubility product of silver chromate is 9×10^{-12}, so precipitation of Ag_2CrO_4 starts when $[Ag^+]^2[CrO_4]$ reaches this value:

$$K_{sp} = 9 \times 10^{-12} = (0.1)^2 \; [CrO_4^{2-}]$$

$$[CrO_4^{2-}] = 9 \times 10^{-10}$$

The solubility product K_{sp} for barium chromate is 2×10^{-10}:

$$K_{sp} = 2 \times 10^{-10} = (0.20)[CrO_4^{2-}]$$

$$[CrO_4^{2-}] = 10^{-8}$$

Thus a smaller concentration of chromate ion is necessary for precipitation of Ag_2CrO_4. Silver precipitates first. When barium begins to precipitate, $[CrO_4^{2-}] = 10^{-8}$. Using this value in the equation for the solubility product of silver chromate, we obtain

$$9 \times 10^{-12} = [Ag^{2+}](10^{-8})$$

$$[Ag^+] = 3 \times 10^{-2}$$

Originally the silver ion concentration was 0.1 M; when barium first precipitated, the silver ion concentration was 0.03 M. Thus 70 percent of the total silver was precipitated before barium precipitation began.

The solubility product equation in terms of concentration applies to very dilute solutions. For example, it applies when solutes of very difficult solubility are considered and very few foreign ions are present. This equation does not apply to moderately soluble electrolytes or for poorly soluble electrolytes in a solution having a large concentration of other ions. The foreign ions do not need to be the same ions in common in order to cause changes in solubility.

Solubility can be expressed in terms of activity:

$$K_{sp,a} = a_A a_B = \gamma_A C_A \ \gamma_B C_B = \gamma_A \gamma_B \ K_{sp}$$

where $K_{sp,a}$ is the thermodynamic solubility product.

These solubility product considerations are applicable to almost all dilute water systems at or near the surface of the earth—lakes, streams, and groundwater (where sorption–desorption energy of the solid materials is negligible).

ACID–BASE EQUILIBRIA

The Lowry–Bronsted definition of acids and bases is that an acid is a proton donor and a base is a proton acceptor. Some chemical species may act as either acids or bases, depending on the circumstances. They are termed "amphoteric." Water is an example:

$$H_2O + H^+ \leftrightharpoons H_3O^+$$
$$H_2O \leftrightharpoons H^+ + OH^-$$

When a molecule or ion loses a proton, it forms a second species that is able, because of the very manner of its formation, to take back a proton, and is therefore a potential base. This resulting base is termed the "base conjugate to the original acid." An example is formic acid, HCOOH.

$$HCOOH + H_2O \leftrightharpoons H_3O^+ + COOH^-$$
$$\text{acid} \qquad\qquad \text{base conjugate}$$

Likewise, there is an acid conjugate to every base. The two members of any such couple are termed a "conjugate acid–base pair." The charge is not important except in that the base is always one unit more negative than the acid. An example is the pair HSO_4^- and SO_4^{2-}.

ACIDITY

In aqueous solutions there can be no free protons. They will always hydrate to form a hydronium ion, H_3O^+. The hydronium ion is the strongest acid that can be present in any aqueous solution. Any stronger acid will lose protons to the water solvent:

$$HCl + H_2O \leftrightharpoons H_3O^+ + Cl^-$$

It is customary and convenient to express the acidity or proton-donating tendency of a solution in terms of the hydrogen ion activity.

Acidity or pH is defined as

$$pH = -\log[H^+]$$

where pH is the negative logarithm of the hydrogen ion activity. If the hydrogen ion activity is 10^{-10}, the pH is 10.

Experimental measurements of pH determine the effective concentration rather than the actual concentration because of reduced H^+ activity by hydration. Only for extremely dilute solutions is the measured pH a measure of the actual concentration.

WEAK ELECTROLYTE EQUILIBRIA

Acids that are weak electrolytes in water enter into equilibrium with the solvent. The general acid is designated "AH." It loses protons reversibly to the water molecules:

$$AH + H_2O \rightleftharpoons A^- + H_3O^+$$

The equilibrium constant is

$$K = \frac{[A^-][H_3O^+]}{[AH][H_2O]}$$

In dilute solutions the water concentration is considered to be constant at 55.51 mol L^{-1}. The equation rearranges to

$$55.51 = \frac{[A^-][H_3O^+]}{[AH]} = K_a$$

where K_a is the acid dissociation constant.

Example: Formic acid, HCOOH:

$$HCOOH + H_2O \rightleftharpoons HCOO^- + H_3O^+$$

The dissociation constant is 1.8×10^{-4}.

What is the hydrogen ion concentration in a 0.1 molar solution?

If the hydrogen ion concentration is designated y, the formate ion concentration is also y, and HCOOH = $0.1 - y$:

$$1.8 \times 10^{-4} = \frac{(Y)(Y)}{0.1 - y} \qquad y = 4.2 \times 10^{-3}$$

The concentrations of H_3O^+ and $HCOO^-$ are each 0.0042. The concentration of HCOOH is $0.1 - 0.0042$. In other words, the amount of dissociation is negligible.

What happens if, for purposes of calculation, we use 0.1 as the concentration of HCOOH, rather than $0.1 - y$:

$$1.8 \times 10^{-4} = \frac{y^2}{0.1} \qquad y = 4.2 \times 10^{-3}$$

The same result is obtained. When the ratio of the dissociation constant to the concentration is 10^{-3} or less, the approximation may be used:

$$K_a = \frac{[A^+][H_3O^+]}{[AH]}$$

where $[A^+]$ equals $[H_3O^+]$ and $[AH]$ is the concentration, C, of the acid.

$$K_a = \frac{[H_3O^+]^2}{C}$$

$$[H_3O^+] = \sqrt{K_a C}$$

The negative log of K_a is designated "pK_a" by analogy with pH. Substitution and taking the log of both sides of the above equation,

$$pH = \tfrac{1}{2}pK_a - \tfrac{1}{2}\log C$$

The solute species does not have to be a neutral molecule. NH_4Cl dissociates to $NH_4^+ + Cl^-$. The ammonium ions enter into an equilibrium with water.

$$NH_4^+ + H_2O \rightleftharpoons NH_3 + H_3O^+$$

$$K_a = 5.7 \times 10^{-10}$$

Compounds that accept protons from water dissociate as weak bases. Some examples are

$$NH_3 + H_2O \rightleftharpoons NH_4^+ + OH^-$$

$$H_2PO_4^- + H_2O \rightleftharpoons H_3PO_4 + OH^-$$

The generalized equation for the equilibrium constant is

$$K_b = \frac{[\text{conjugate acid}][\text{OH}^-]}{[\text{base}]}$$ (for a dilute solution in which $[\text{H}_2\text{O}] = 55.51$)

[Conjugate acid] = [OH$^-$] because of the nature of their formation. As in the case of acids, we can make an approximation for [Base], on the assumption that the amount of dissociation is so small that the loss of base by dissociation is negligible in comparison with the original concentration of the base.

If C is the concentration of the base, we can follow the same reasoning applied above to weak acids and derive

$$[\text{OH}^-] = K_b C$$

$$\text{pOH} = \tfrac{1}{2}\text{p}K_b = \tfrac{1}{2}\log C$$

(There are tables of K_a and pK_a values for acids and K_b and pK_b values for bases.)

For any given weak acid or weak base, if the concentration is specified, the acidity at standard conditions can be determined by looking up the dissociation constant in a table and substituting the two values into the equation. As pOH is directly related to pH, usually only pH is specified.

Water is a weak electrolyte.

$$\text{H}_2\text{O} \rightleftharpoons \text{H}^+ + \text{OH}^-$$

$$K = \frac{[\text{H}^+][\text{OH}^-]}{[\text{H}_2\text{O}]}$$

Because [H$_2$O] is considered unity (standard state), the conventional ion product constant for water is

$$K_w = [\text{H}^+][\text{OH}^-]$$

At temperatures near room temperature, the value is very close to 1×10^{-14}.

$$10^{-14} = [\text{H}^+][\text{OH}^-]$$

$$\text{pH} + \text{pOH} = 14$$

If pOH is calculated for a certain K_b, the pH is immediately known from the above relationship. The above equation is also the basis for the pH scale. When water is neutral, there is no excess of H^+ or OH^-. $[H^+] = [OH^-]$ or pH = pOH.

Therefore, $2pH = 14$; pH = 7.

EQUILIBRIUM FOR A CONJUGATE PAIR

When an acid dissociates, it gives up a proton to form a conjugate base. The conjugate base can also react with water, taking up a proton, to once again form the acid. The two dissociation constants must be related, because both depend on the ability of the conjugate base to hold the proton.

Consider an acid HA and the conjugate base A^-. For the acid

$$HA \rightleftharpoons H^+ + A^-$$

$$K_a = \frac{[H^+][A^-]}{[HA]}$$

For the conjugate base

$$A^- + H_2O \rightleftharpoons HA + OH^-$$

$$K_b = \frac{[HA][OH^-]}{[A^-]}$$

Multiply the top and bottom (numerator and denominator) by H^+:

$$K_b = \frac{[HA][OH^-][H^+]}{[A^-][H^+]} = \frac{[HA]}{[A^-][H^+]} \times [OH^-][H^+]$$

But

$$[OH^-][H^+] = K_w \quad \text{and} \quad \frac{[HA]}{[A^-][H^+]} = \frac{1}{K_a}$$

Therefore

$$K_b = \frac{K_w}{K_a} \quad \text{and} \quad K_a K_b = K_w$$

Remember that K_w is a constant for dilute solutions. Thus a strong base is always conjugate to a weak acid, and a weak base is always conjugate to a strong acid.

  Compounds such as H_2CO_3 or H_3PO_4 contain more than one acid hydrogen, and anions such as CO_3^{2-} or PO_4^{3-} that are capable of adding more than one proton are termed "polyprotic acids and bases." Ionization occurs in a stepwise fashion, and the constants for successive steps are termed K_1, K_2, K_3, etc.

$$H_2CO_3 \rightleftharpoons H^+ + HCO_3^- \qquad K_1 = 4.5 \times 10^{-7}$$

$$HCO_3^- \rightleftharpoons H^+ + CO_3^{2-} \qquad K_2 = 5.6 \times 10^{-11}$$

where K_1 and K_2 are the acid dissociation constants for H_2CO_3 and HCO_3^-. The successive dissociation constants are always smaller, for it is always more difficult for the next step to occur.

  If the constants for a polyprotic acid differ by several orders of magnitude, as they usually do, only the first ionization contributes appreciably to the pH. For the example above, consider a 0.1-molar solution of H_2CO_3:

$$K_1 = 4.5 \times 10^{-7} = \frac{[H^+][HCO_3^-]}{[H_2CO_3]}$$

Because $[H^+] = [HCO_3^-]$:

$$[H^+] = \frac{[H^+]^2}{(0.1 - [H^+])}$$

Because $[H^+] \ll 0.1$:

$$[H^+]^2 = (0.1)(4.5 \times 10^{-7}) = \frac{[H^+]^2}{0.1}$$

$$[H^+] = 2.1 \times 10^{-4}$$

For the second dissociation:

$$K_2 = 5.6 \times 10^{-11} = \frac{[H^+][CO_3^{2-}]}{[HCO_3^-]}$$

  Because the second ionization is much weaker, $[H^+]$ is not changed appreciably by the second ionization. The $[HCO_3^-]$ is also not considered to change appreciably. These two values, which are equal, cancel out in the equation. Therefore

$$[CO_3^{2-}] = 5.6 \times 10^{-11} \text{ mol L}^{-1}$$

The final results are

$$[H_2CO_3] = 0.1 \text{ mol } L^{-1} - 0.00021$$

$$[HCO_3^-] = 0.00021 - 0.000000000056$$

$$[H^+] = 0.00021 + 0.000000000056$$

$$[CO_3^{2-}] = 0.000000000056 \text{ mol } L^{-1}$$

Only the first ionization contributed appreciably to the pH.

AMPHOTERIC SPECIES

Species such as $H_2PO_4^-$ and HCO_3^- may lose or gain a proton. What is the acidity of a solution containing such a species?

$$RH \rightleftharpoons R^- + H^+$$

$$RH + H_2O \rightleftharpoons RH_2^+ + OH^-$$

The equilibrium constants are the acid and base dissociation constants for the species:

$$K_a = \frac{[R^-][H^+]}{[RH]} \qquad K_b = \frac{[RH_2^+][OH^-]}{[RH]}$$

Regardless of what dissociations occur, the overall electrical balance is maintained:

$$[H^+] + [RH_2^+] = [OH^-] + [R^-]$$

It is now possible to make a series of substitutions into the above equations in order to put everything in terms of $[H^+]$:

$$K_w = [OH^-][H^+] \qquad \therefore \qquad [OH^-] = \frac{K_w}{[H^+]}$$

Substitute into K_b equation:

$$[RH_2^+] = \frac{K_b[RH]}{[OH^-]} = \frac{K_b[H^+][RH]}{K_w}$$

Rearrange K_a equation:

$$[R^-] = \frac{K_a[RH]}{[H^+]}$$

Substitute both into electrical balance equation:

$$[H^+] + \frac{K_b[H^+][RH]}{K_w} = \frac{K_w}{[H^+]} + \frac{K_a[RH]}{[H^+]}$$

Multiply both sides by $[H^+]$:

$$[H^+]^2 + \frac{K_b[H^+]^2[RH]}{K_w} = K_w + K_a[RH]$$

$$[H^+]^2 \frac{(1 + K_b[RH])}{K_w} = K_w + K_a[RH]$$

$$[H^+] = \sqrt{\frac{K_w + K_a[RH]}{1 + (K_b[RH]/K_w)}}$$

If K_w is much smaller than $K_a[RH]$ or $K_b[RH]$, then the equation can be reduced further. The reason is that in this case, K_w added to the $K_a[RH]$ term in the numerator is negligible and may be dropped. In the denominator, the term containing K_w will be so large that "1" added to it is negligible, and the "1" may be dropped. The equation reduces to

$$[H^+] = \sqrt{\frac{K_a[RH]}{K_b[RH]/K_w}} = \sqrt{\frac{K_a K_w}{K_b}}$$

This shows the hydronium concentration to be dependent on only the dissociated constituents. Therefore, the pH is independent of the amphoteric concentration. H_2CO_3, for example, has a fixed pH that does not vary even though more H_2CO_3 is added to the solution. The approximations used in developing the equation hold for many amphoteric species.

OXIDATION–REDUCTION

The Lewis definition of an acid is general. An "acid" is an electron pair acceptor. A "base" is an electron pair donor. The processes of electron transfer in a solution are termed oxidation–reduction reactions. The Eh, or oxidation potential, can be considered to be the measure of the availability of electrons in the system, in the same way that pH is a measure of the proton availability. Just as acid–base equilibria are treated in pairs—an acid and the conjugate base, and vice versa—oxidation–reduction equilibria are treated in pairs: the reduced form and the oxidized form. The oxidized form is that which has lost electrons relative to the reduced form:

$$Fe^{3+} + \tfrac{1}{2}H_2 = Fe^{2+} + H^+$$

The two half-reactions are

$$Fe^{3+} + e^- = Fe^{2+}$$

$$\tfrac{1}{2}H_2 = H^+ + e^-$$

The chemical redox (reduction–oxidation) reaction can produce an electric current in some cases, that is converting chemical energy into electric energy. The voltage of the electric current produced is proportional to the energy of the chemical redox reaction. Electrical voltages can be measured very precisely, and provide a way of measuring the oxidation–reduction reaction energy.

Substances differ in their ability to combine with electrons and become reduced or to lose electrons and become oxidized. The electrode potential (E°) is a measure of its oxidation–reduction ability. The potential of a single electrode cannot be determined. However, the difference in potential of two electrodes can be measured. When the difference in potential is compared to the reference standard, relative comparisons between half-cells may be made. The universally accepted standard reference is the hydrogen electrode (the voltage has been arbitrarily set to 0.000 V):

$$
\begin{array}{ll}
Fe^{3+} + e^- = Fe^{2+} & E^\circ = +0.771V \\
\underline{\tfrac{1}{2}H_2 = H^+ + e^-} & \underline{E^\circ = 0.000} \\
Fe^{3+} + \tfrac{1}{2}H_2 = Fe^{2+} + H^+ & E^\circ = +0.771V
\end{array}
$$

A standard electrode potential E° is the potential of an electrode when each substance involved in the half-reaction is in its standard state. The standard state of an ion in solution is at unit activity. An elemental metal is in its standard state. The standard state of a gas is at 1-atm pressure. The more positive the potential of the half-reaction, the greater the oxidizing power of the oxidized form of the redox couple. Conversely, the more negative the half reaction potential, the greater the reducing power of the reduced form of the redox couple.

The electrode potential is different from its standard potential if the redox substances are not in their standard states. The electrode potential changes as the concentrations of the soluble species change and may be calculated with the Nernst equation.

For the half-reaction

$$A_{ox} + n_{e-} = A_{red}$$

the Nernst equation is

$$E = E° + 2.303 \frac{RT}{nF} \log \frac{[A_{ox}]}{[A_{red}]}$$

where E = potential in volts; $E°$ = standard electrode potential; R = gas constant, 8.314 joules per degree-mole; T = absolute temperature, 273 + temperature in degrees centigrade; n = number of electrons involved; F = Faraday constant, 96,500 coulombs (C); A_{ox} = molar concentration of the oxidized form; and A_{red} = molar concentration of the reduced form.

The Nernst equation can be derived as follows. The chemical potential of a species u_i is defined by the equation

$$u_i = RT \ln a_i + u_i$$

where a_i is the activity of an ion compared with some arbitrary standard state having a chemical potential $u_i°$.

For the reaction

$$Ox + n_{e-} = red$$

$$u_{red} = RT \ln A_{red} + u_{red}°$$

$$u_{ox} = RT \ln A_{ox} + u_{ox}°$$

The difference in chemical potential from the initial to the final state is $u_{red} - u_{ox}$ or ΔG, where ΔG is the change in free energy.

$$\Delta G = RT \ln A_{red} - RT \ln A_{ox} + G°$$

where $\Delta G = nFE$ and $G° = -nFE°$, where n is the number of electrons involved, F is the Faraday constant, and E is the potential in volts.

At ordinary room temperature, 25°C, the Nernst equation has the form

$$E = E° + \frac{0.059}{n} \log \frac{[A_{ox}]}{[A_{red}]}$$

The electrode potential depends on the ratio of oxidized to reduced forms in solution.

The endpoint in an oxidation–reduction titration can be detected by the color change of a visual indicator, or the titration curve can be plotted from potentiometric measurements. The potentiometric titration is the difference between an indicator electrode and a reference electrode.

Suggested Readings

American Public Health Association, 1985. "Standard Methods for the Examination of Water and Wastewaters." 16th Ed. Published by the Amer. Public Health Assoc., Amer. Water Works Assoc., and Water Poll. Control Fed., Washington, D.C. 1268 pp.

Bormann, F.H., and G.E. Likens, 1981. "Pattern and Process in a Forested Ecosystem." Springer-Verlag, New York. 253 pp.

Drever, J.I., 1982 "The Geochemistry of Natural Waters." Prentice-Hall, Inc., Englewood Cliffs, N.J. 388 pp.

Hem, J.D., 1985. (3rd Edition). *Study and interpretation of the chemical characteristics of natural water.* U.S. Geological Survey Water Supply Paper 2254. Washington, D.C. 263 pp.

Novotny, V., and G. Chesters, 1981. "Handbook of Nonpoint Pollution: Sources and Management." Van Nostrand Reinhold Co., New York. 555 pp.

Reeder, S.W., B. Hitchon, and A.A. Levinson, 1972. Hydrogeometry of the surface waters of the Mackenzie River drainage basin, Canada: I. Factors controlling inorganic composition. *Geochimica et Cosmochimica Acta* **36**: 825–865.

Reynolds, R.C., and N.M. Johnson, 1972. Chemical weathering in the temperate glacial environment of the Northern Cascade Mountains. *Geochimica et Cosmochimica Acta* **36**: 537–544.

Thurman, E.M., 1985. "Organic Geochemistry of Natural Waters." Nijhoff/Junk Publishers, Dordrecht, Netherlands. (Distributed by Kluwer Academic Publishers, Boston.). 497 pp.

Chapter 2

Sampling Procedures

Stream Discharge Measurement

Stream discharge is probably the most important parameter measurement in wildland water quality. Discharge (Q) is a measure of the volume of water passing a given point on the stream per unit period of time. The common English unit is cubic feet per second (cfs), and the metric expression is cubic meter per second (cms) or liters per second (Ls^{-1}), if stream discharge is less than 1 cms.

Stream discharge can be measured by a number of different methods. Only the velocity area method will be discussed here.

The velocity area method is based on the continuity equation:

$$Q = Av$$

where Q is discharge in cubic feet per second or cubic meters per second, A is cross-sectional area normal to the mean velocity in square feet (ft^2) or square meters (m^2), while v is the mean velocity of the cross section in feet per second or meters per second. Discharge is measured this way for a number of subsections of the stream cross-sectional area, the sum of these being the total discharge.

Figure 2.1. Stream velocity determinations can be made with a current meter and a top-setting wading rod. The current meter can quickly be placed at 0.6 depth after reading the stream's depth on the wading rod.

MEASUREMENT OF CROSS-SECTIONAL AREA

The stream top width is subdivided into a number of subsections depending on the degree of variability of the depth and velocity across the stream and the degree of precision required. If high precision is desired, more subsections are used. The depth of each subsection is determined with a wading rod (see Figure 2.1) or some other measuring device. A rule of thumb is no more than 10 percent of the flow should occur in any one section.

MEASUREMENT OF VELOCITY WITH CURRENT METER

The depth of water is measured for each section, and the velocity observation depth of the meter is decided from that depth. If the depth is less than 0.6 m (2 ft), the meter is set at 0.6 of the depth from the surface. When depths of greater than 0.6 m are encountered, the average of the measurement at 0.2 and 0.8 of the depth from the

Figure 2.2. Stream velocity determinations can be measured by vertical-axis current meters. The Price meter (left) or the Gurley Pygmy meter (right) may be used depending on the stream depth and stream velocity. Each has a separate rating curve.

surface is the mean velocity. The velocity observation is made in the center of the cross-section area subsection. Most current meters are now direct-reading; however, the user should be familiar with the specific instrument used. Examples are shown in Figure 2.2.

Water Quality Sampling[1]

Chemical elements in various forms are distributed throughout the hydrologic environment—in the atmosphere, precipitation, surface water, soil water, and groundwater. Measurements in waters are most important, and the validity of each measurement begins on sampling. Irrespective of scrutiny and quality control applied in performing laboratory analyses, reported data are no better than the confidence that can be placed in the representativeness of the sampling.

The term "representative" means that the sample represents all possible samples within the population. To achieve representativeness for data comparability and consistency, it is necessary to minimize, or at least standardize, sampling bias as it relates to site selection, sampling frequency, sample collection, sampling devices, and sample handling, preservation, and identification.

SITE SELECTION

Whether the basic objective of the monitoring program is reconnaissance, long-term trend evaluation, or solving specific problems, the first step in any study is to define the hydrologic boundaries and then establish the sampling site locations above, within, or below these boundaries. Most of the sampling points are linked to specific hydrologic processes, for example, precipitation.

Precipitation

Several factors should be considered when selecting a sampling site: study objectives, accessibility, physical characteristics, sources

---

[1] Adapted from U.S. Geological Survey 1977, "National Handbook of Recommended Methods for Water-Data Acquisition," Chapter 5, Chemical and Physical Quality of Water and Sediment.

of contamination, and personnel and facilities available for conducting the study.

The objectives of wet and dry atmospheric precipitation studies may be as diverse as their number; however, most studies are designed to quantify the temporal and spatial variation of atmospheric precipitation quality as it may relate to various human activities and natural events.

In studies designed to quantify regional precipitation quality or to establish long-term trends, the most important factor in site selection is avoidance or minimization of local sources of contamination. However, if the study objective is to determine the relationship between a specific source (or group of sources) and precipitation quality, site selection becomes far more complex. In this case, the objective requires that the site be chosen to optimize the impact of a specific source.

If a region is typified by a certain type of land use, the sampling site should reflect this usage. For example, if regional land use is dominated by agricultural activity, then locating a sampling site within or adjacent to or downwind of that activity is not precluded. Conversely, if wilderness dominates regional land use, then steps should be taken to avoid potential sources of anthropogenic contamination.

Where regional quality of precipitation is the study objective, the following criteria are suggested in order to minimize undesirable impacts:

1. Sources of contamination from air, ground, or water traffic should be avoided.
2. Surface storage of agricultural products, fuels, or other foreign materials should be avoided.
3. Continuous sources of contamination within 50 km upwind of the site and 30 km in other directions should be avoided.
4. Samplers should be installed over undisturbed land.

Surface Water

Many factors are involved in the proper selection of surface-water sampling sites, including the following: study objectives; accessibility; flow, mixing, and other physical characteristics of the waterbody; point and diffuse sources of contamination; and personnel and facilities available to conduct the study. In the case of dissolved

Sampling bedload with a cable-mounted Helley–Smith bedload sampler from a cable car. Indian River, Southeast Alaska.

constituents, dispersion depends on the vertical and lateral mixing within the cross section of a body or water. The hydrologist collecting the samples must, therefore, not only be familiar with the mixing characteristics of streams and lakes but must also have a good understanding of the role of fluvial-sediment transport, deposition, and chemical sorption phenomena. Therefore, the study can usually be tied to the physiographic features of the area under consideration. In most cases, the quality of water flowing into the lake or past a particular point along a stream can be related directly to inputs within the drainage basin, including the entire drainage area upstream of a selected point.

Accessibility to any sampling site is a requirement that is often directly related to sampling costs. Bridges are normally the first choice for locating a stream sampling station because they provide ready access and also permit sampling at any point across the stream. Sampling locations on lakes and reservoirs, as well as those on larger rivers, may require the use of a boat. Frequently, however, a boat will not only take more time in traversing a waterbody but also will make the manipulation of the sampling equipment difficult.

The ideal sampling situation for a stream is a cross section that would yield the same concentrations of all constituents from all points along the cross section, and a sample taken at any time would yield the same concentrations as one taken at any other time. This situation never exists in nature for any length of time, a fact that reinforces the need for careful site selection in order to ensure, as nearly as possible, that samples are taken where there is uniform flow and good mixing conditions.

The availability of streamflow discharge records can be an important consideration in the selection of sampling sites on streams. Adequate discharge data are essential for estimating the total loads carried by the stream past the sampling point. If a gaging station is not conveniently located on a selected stream, the hydrologist should expore the possibility of obtaining discharge data by direct or indirect means.

Groundwater

Many factors are involved in proper selection of groundwater sampling sites, including the following: study objectives; location, number, and spacing of wells; vertical location of screen material; type of well casing and screen; how the water sample will be taken;

at what depth the water sample will be drawn; and installation of well to minimize water column contamination.

SAMPLING FREQUENCY

Precipitation

The frequency of sampling atmospheric waters is largely determined by the study of objectives. Source-related studies often necessitate short-term event sampling, whereas long-term trends and averages can be determined by weekly, biweekly, or even monthly sampling. In the United States, the National Atmospheric Deposition Program (NADP) is currently collecting and analyzing wet precipitation samples on a weekly basis; the dry precipitation samples are collected at a weekly frequency, and are composited and analyzed every 2 months.

Surface Water

Surface-water monitoring, which consists of repetitive, continuing measurements to define variations and trends at a given location, should include collection of water samples at least monthly, with particular emphasis on extreme events. Stream bottom sediment samples should be collected at least yearly from fresh deposits, and preferably during both spring and fall seasons, particularly when water quality analyses include metals. Additional sampling related to other flow conditions may be desirable.

Groundwater

There is no fixed answer for sampling frequency of groundwater for water quality. The water sample needs to be representative of groundwater quality, and cannot be contaminated by sampling or handling procedures. Because of hydraulic gradients and variable chemical processes, the water quality in or near the well casing is probably not representative of the the overall groundwater quality. Sampling frequencies for groundwater have been determined from hazard ratings based on the most conservative constituent or the constituent of the most interest. Such hazard ratings are based on the current constituent concentration and the tolerated dose or concentration. These rating formats presuppose a normal distribution of constituent concentration values and a known tolerance concentration.

The alternative method for groundwater sampling frequency is based on the average groundwater velocity. If the groundwater velocity is less than 23 m per year, one sample year is considered adequate to characterize the water quality. As groundwater velocities increase, the number of samples also increases.

Evaluation of the variability in available water quality data must precede any decision as to the number of samples and collection frequency required to maintain an effective monitoring program.

SAMPLE COLLECTION

Precipitation

Precipitation samples are usually collected in either cylindrical glass or plastic containers. Plastic containers are used if inorganic analyses are desired, and glass containers are used for organic constituents. Sampling containers may be continually exposed to collect an integrated wet and dry sample (bulk precipitation collectors) or may be preferentially exposed to allow separation of the wet and dry components (wet/dry precipitation collectors).

Surface Water

Sample collection in streams and rivers varies from the simplest of hand-sampling procedures at a single point to the more sophisticated multipoint sampling techniques known as the "equal-discharge-increment (EDI) method" or the "equal-width-increment (EWI) method," formerly described as the "equal-transit-rate (ETR) method" [U.S. Geological Survey (USGS) techniques].

Generally, the number and type of samples to be taken depends on the width, depth, and stream discharge. The larger the number of individual points sampled, the more nearly will the composite sample represent the water body.

Any sample container should be rinsed with the stream water to be sampled. The sample container or bottle and cap should be filled at the sampling point and then emptied three times (see Figure 2.3). The easiest approach to stream sampling is to face upstream and sample upstream. If rinsing the cap and bottle, turn to the side or downstream to empty the bottle to avoid disturbing the stream bottom material or to avoid possible contamination of the water column.

Figure 2.3. Surface water quality sample containers should be rinsed at least three times with the water to be sampled. A common sample container is the 1-L narrow-mouthed, cross-linked polyethylene, low-density polyethylene, or high-density polyethylene containers made of nalgene.

Water is often sampled by filling a container held just beneath the surface of a body of water, commonly referred to as a "dip" or "grab sample" (see example in Figure 2.4). Published data and discussions with investigators reveal that a high percentage of samples are obtained in this manner. Using a weighted-bottle holder that allows a sample bottle to be lowered to any desired depth, opened for filling, closed, and returned to the surface, improves on this method. If an open bottle is lowered to the bottom and raised to the surface at a uniform rate and at such a rate as to have the bottle just filled on reaching the surface, the resulting sample will roughly approach the collection of what is known as a "depth-integrated sample."

Because many organic compounds are only slightly water-soluble, they tend to be absorbed by particulate matter in the water. This requires a sampling method that collects a representative amount of suspended material and one that does not allow transfer of water

Figure 2.4. Water quality sampling in shallow streams may be done best by grab samples. The sample bottle is rinsed with the water to be sampled at least three times. If a sample preservative is used, the preservative should be added first, and the container not rinsed.

from the sampling container without transfer of an appropriate amount of that suspended material.

It is recommended that all stream samples be collected by depth-integrating methods using either a hand-held or a cable-and-reel-suspended sampler whenever it is practical to do so. The only exception might be in the case of shallow streams where the depth is insufficient to allow true depth integration. In such cases, dip samples collected at one or more verticals across the stream are appropriate; however, the sample container should be held carefully just beneath the water surface in order to avoid disturbing the streambed.

The number of verticals to be sampled is usually decided by the person actually collecting the sample. For small streams, a depth-integrated sample taken at the center of flow is usually adequate. Larger streams require sampling of several verticals at centroids of equal discharge. This method requires knowledge of the streamflow distribution in the cross section before sampling verticals can be selected.

The number of verticals required depends on the streamflow and sediment characteristics at the time of sampling as well as the desired accuracy of sampling. For all except very narrow (< 1 m) and shallow (< 0.1 m) streams, 10 verticals are usually sufficient.

Obviously, the multipoint sampling techniques can become very time-consuming and, consequently, expensive. An alternate method often used consists of sampling at the quarter points or other equal intervals across the width of the stream. Composite samples obtained from individual samples collected at the quarter points can be fairly representative, providing that the stream cross section is properly located.

Automatic pumping water samplers are often used for sequential sampling or programmed sampling in remote areas. The sample intake should be located in the stream cross section where the most mixing occurs or simply in the stream center. The automatic pumping sampler water quality analyses should be compared to analyses from depth-integrated samples taken at the same location and time.

Lakes and Reservoirs

The number of sampling sites on a lake or reservoir will vary with the size and shape of the basin. In shallow lakes having a circular basin, a single site in the deepest part of the lake may be sufficient to describe the distribution and abundance of the constituents in solution. In natural lakes, the deepest point is often near the lake center, and in reservoirs, the deepest area is near the dam. In lakes with irregular shape and with several arms and bays that are protected from the wind, additional sampling sites may be needed to define the water quality adequately.

Many lake measurements are now made in situ with the use of sensor probes and automatic readout or recording devices. Single- and multiparameter instruments are available for measuring temperature, depth, pH, specific conductance, dissolved oxygen, some cations and anions, and light penetration. Sensitivity and calibration of the chemical sensors are limiting factors in their use.

Lake water samples may be collected at any desired depth with a

Kemmerer-type sampler and brought to the surface for filtering or other preanalysis treatment. Nonuniform vertical mixing of chemical constituents often occurs because of wind and temperature changes, the shape of the lake basin, biological activity, and many other factors. Therefore, water samples should be collected from several depths in the water column.

Groundwater

Groundwater sample collections should be of the groundwater and not the water standing in the casing. It is desirable to have a well pumped or bailed until the well is thoroughly flushed of standing water. A suggested minimum is 4–10-bore volumes. Another approach is to monitor the water level until stabilized while pumping, or monitor water pH and temperature while pumping. Sample withdrawal may be by bailer, suction lift, submersible pump, air lift, persitolic pump, or multilevel syringe samplers. In any case, the selection of sampling equipment will be dependent on the nature of parameters of interest.

SAMPLING DEVICES AND CONTAINERS

Precipitation

Glass or nalgene containers, located 1–2 m above ground, are recommended for the collection of bulk precipitation samples. Where evaporative loss presents a problem, funnel collectors may be used. In areas where precipitation is light or where snowfall is collected, large containers are needed to collect an amount necessary for subsequent analysis. Bulk precipitation devices, by their very nature, are often subject to extensive sample contamination by insects and droppings; however, they are simple to use and sufficient for most studies. Bulk precipitation collection devices do not usually require power unless sequential samples are required.

Glass or nalgene containers, located 1–2 m above ground, are also used for collecting wet and dry precipitation samples. To overcome problems of evaporative loss and to minimize wet sample contamination, sampling equipment has been designed to selectively seal either the wet or dry sampling bucket when the other is in use. Under dry conditions, the dry bucket is exposed while the wet bucket is sealed by the cover. During a rain or snow event, a conductivity sensor causes the wet bucket to be exposed and the dry bucket to be covered. Following the wet precipitation event, the dry bucket is again exposed. These collectors require electrical power and may be operated with direct or alternating current. The size of the sample bucket or bottle will depend on the volume of sample required for analysis and the expected amount of precipitation occurring over the sampling period. When possible, a type of collection device comparable to those used in the existing network should be chosen for a project in order to facilitate network operation and yield comparative data for equal validity.

Surface Water

To overcome problems associated with collecting representative samples, equipment that has been specifically designed and thoroughly tested is favored. Several depth-integrating samplers are available and suitable for this purpose. In streams that can be waded, the US DH-48 suspended-sediment sampler can be used successfully.

The US DH-59 suspended-sediment sampler was designed to be suspended by a hand-held rope in streams too deep to be waded. Both of these samplers are simple and of straightforward design.

Thoroughly cleansed linear nalgene or glass bottles fitted with screwcaps may be used for water samples collected with depth-integrating samplers or from precipitation samplers. Nalgene containers are generally preferred for inorganic samples, and glass is preferred for organic samples. Stream bottom sediments can be placed in clean, wide-mouthed plastic or glass bottles with Teflon-lined caps, depending on whether inorganic or organic analyses will be performed. Nalgene bottles must not be used for organic samples and certain trace metals because it is known that they introduce interferences and have sorption characteristics. To avoid problems in transfer, most water samples for organic analyses should remain in the collection container until analysis is begun because the laboratory procedure often includes a method for rinsing the sample container with the extraction solvent. A sufficient volume of sample should be collected to satisfy the requirements of each analysis and also to permit analysis of duplicate and fortified samples. Breakage of glass sample bottles can usually be overcome by shipping them in expanded polystyrene molds that fit the bottles.

Groundwater

Considerations for sample containers for groundwater are the same as those for surface water samples. An additional consideration, however, is to recognize potential chemical interferences from the well casing, well screen, or sampling device.

SAMPLE HANDLING

Deteriorated samples negate all the efforts and cost expended in obtaining good samples. In general, the shorter the time that elapses between the collection of a sample and its analysis, the more reliable will be the analytical results. For certain constituents and physical values, immediate analysis in the field is required in order to obtain reliable results because the composition of the sample almost certainly will change before it arrives at the laboratory. However, some samples can be satisfactorily preserved by chilling or by adding suitable acid or germicide or by other special treatment. They may then be allowed to stand for a longer period of time before analysis.

Precipitation

Preferably, pH and conductivity measurements of precipitation should be made in the field and also immediately on receipt of the samples in the laboratory. Following pH and conductivity determinations in the laboratory (deionized water is added to leach dry samples), the remaining sample should be filtered through a 0.45 μm polycarbonate membrane or Teflon filter directly into the sample storage bottle. This filter size is used to separate total from dissolved constituents. In some cases, the filters may be stored and retrieved for subsequent analysis of the residue. Samples should be stored at 4°C in an unlighted environment until analyses are performed. No preservatives should be used. Because of the eventual loss of nitrate, phosphate, and ammonia, it is advisable to analyze all samples as soon as possible after collection.

Surface Water and Groundwater

Determination of temperature, pH, specific conductance, and dissolved gases should be made in the field. Samples for metal analysis can be preserved by the addition of nitric acid; samples for organic constituent determinations preserved by chilling or freezing; and samples for the determination of such biodegradable substances as nitrates, phosphates, and surfactants by chilling the sample immediately and storing the sample in the dark at a temperature just above freezing until the analyses are made. It is necessary, however, to select the method of analysis and determine what preservative is recommended for that particular determination before adding a preservative to any sample because some preservatives interfere with analytical measurements. Samples for the determination of certain dissolved inorganic and organic species must not be frozen inasmuch as it is not always possible to reconstitute the original sample exactly as it was before freezing.

The Resource Conservation and Recovery Act (RCRA) requires that the transfer and management of a chemical sample be documented. Determining if a chain of custody is required is

Figure 2.5. The water quality sample container is labeled with permanent ink prior to sampling. The sample label should include the site name, date, time, and name of sampler as a minimum.

usually straightforward. Consider use of the chain of custody procedure if any regulations are being met or legal proceedings are undertaken, or when more than one organization is involved with the sampling, analytical, and data analysis phases of the project.

It is surprising how much data are lost by poorly labeled sample bottles. The sample container should be marked with permanent ink or lab tape on site, prior to sample collection (see Figure 2.5). As a minimum, sample labeling should include sample name and/or number, sample data and time, initials of sampler, and stage–head data or other pertinent information. The same information as well as site observations should be recorded in a fieldbook.

SAMPLE PRESERVATION[2]

Complete and unequivocal preservation of samples of either domestic sewage, industrial wastes, or natural waters, is a practical impossibility. Regardless of the nature of the sample, complete stability for every constituent can never be achieved. At best, preservation techniques can only retard the chemical and biological changes that inevitably continue after the sample is removed from the parent source.

The changes that take place in a sample are either chemical or biological. In the former case, certain changes occur in the chemical structure of the constituents that are a function of physical conditions. Metal cations may precipitate as hydroxides or form complexes with other constituents; cations or anions may change valence states under certain reducing or oxidizing conditions; or other constituents may dissolve or volatilize with temperature changes or with the passage of time. Metal cations may also adsorb onto container surfaces (glass, plastic, quartz). Biological changes taking place in a sample may change the state of an element or a radical to a different state. Soluble constituents may be converted to organically bound material in cell structures, or cell lysis may result in release of cellular materials into solution. The well-known nitrogen and phosphorus cycles are samples of biological influence on sample composition.

---

[2] Adapted from U.S. Environmental Protection Agency, 1974. Methods for Chemical Analysis of Water and Wastes.

TABLE 2.1.

Sample Preservation Methods for Specific Parameters in Water Quality Samples

Parameter	Preservative	Maximum Holding Period
Acidity–Alkalinity	Refrigeration at 4°C	24 hours
Biochemical oxygen demand	Refrigeration at 4°C	6 hours
Chemical oxygen demand	2 mL H_2SO_4 per liter	7 days
Chloride	None required	7 days
Color	4°C	24 hours
Cyanide	NaOH to pH 10	24 hours
Dissolved-oxygen holding	Determine on site	No
Fluoride	None required	7 days
Hardness	None required	7 days
Metals, total	5 mL HNO_3 per liter	6 months
Metals, dissolved	Filtrate: 3 ml 1:1 HNO_3 per liter	6 months
Nitrogen, ammonia	0.8 mL H_2SO_4 per liter—4°C	7 days
Nitrogen, Kjeldahl	0.8 mL H_2SO_4 per liter—4°C	Unstable
Nitrogen, nitrate–nitrite	0.8 mL H_2SO_4 per liter—4°C	7 days
Oil and grease	2 mL H_2SO_4 per liter—4°C	24 hours
Organic carbon	2 mL H_2SO_4 per liter (pH 2)	7 days
pH holding	Determine on site	No
Phenolics	1.0 g $CuSO_4$ per L + H_3PO_4 to pH 4.0–4°C	24 hours
Total phosphorus	2 mL H_2SO_4 per liter	7 days
Orthophosphorus	40 mg $HgCl_2$ per liter—4°C	7 days
Total phosphorus	2 mL H_2SO_4 per liter	7 days
Orthophosphorus	40 mg $HgCL_2$ per liter—4°C	7 days
Solids	4°C	7 days
Specific conductance	4°C	7 days
Sulfate	4°C	7 days
Sulfide	2 mL Zn acetate per liter	7 days
Threshold odor	4°C	24 hours
Turbidity	None available	7 days
Total coliform	4°C	6 hours
Fecal coliform	4°C	6 hours

Methods of preservation are relatively limited and are intended generally to (1) retard biological action, (2) retard hydrolysis of chemical compounds and complexes, and (3) reduce volatility of constituents. Preservation methods are generally limited to pH control, chemical addition, refrigeration, and freezing. Preservation methods increase the allowable holding time (see Table 2.1). Refrigeration at temperatures near 4°C is the best overall preservation technique.

Review

1. Consult the laboratory before collecting and submitting a sample for analysis.
2. Know which specific analyses are to be made on the sample.
3. Collect representative samples.
4. Avoid contaminating the samples once they are collected.
5. Immediately refrigerate samples, or use the proper preservative.
6. Note that refrigeration is the only preservation method to be used on samples taken for biological analysis.
7. If there is doubt about the preservation method to use, store the sample on ice.
8. Label all samples immediately and record them in a fieldbook.
9. List pertinent data on the label or accompanying letter.
10. Keep the time interval between collection and analysis as short as possible.

Suggested Readings

Brinkhoff, H.C., 1977. Continuous on-stream monitoring of water quality. *International Workshop on Instrumentation and Control for Water and Wastewater Treatment and Transport Systems.* Philips Science and Industry Division, London, England.

Brown, E., M.W. Skougstaf, and M.J. Fishman, 1970. Methods for collection and analysis of water samples for dissolved mineral and gases, *In* "Techniques of

Water-Resources Investigations of the United States Geological Survey," Chapter A1, Book 5. U.S. Geological Survey, Washington, D.C.

Buchanan, T.J., and W.P. Somers, 1969. Discharge measurements at gaging stations. In "Techniques of Water-Resources Investigations of the United States Geological Survey," Chapter A8, Book 3. U.S. Geological Survey, Washington. D.C.

Carroll, D., 1962. Rainwater as a chemical agent of geologic processes–a review. U.S. Geological Survey. Water Supply Paper 1535, Washington D.C. 18 pp.

Cavagnaro, D.M., 1977. "Automatic acquisition of water quality data." Volume I: 1970–1975, A Bibliography with Abstracts. NTIS/PS-76/0670. National Technical Information Service, U.S. Department of Commerce, Springfield, Virginia.

Fenn, D., E. Cocozza, J. Isbiter, O. Braids, B. Yare, and P. Roux, 1977. "Procedures manual for groundwater monitoring at solid waste disposal facilities" U.S. Environmental Protection Agency. EPA/530/SW-6111.

Fisher, P.D., and Siebert, J.E., 1977. Integrated automatic water sample collection system. *Journal of the Environmental Engineering Division, American Society of Civil Engineers*, **103**, EE4: 725–728.

Galloway, J.N., and Likens, G.E., 1976. Calibration of collection procedure for the determination of precipitation chemistry. In "First International Symposium on Acid Precipitation and Forest Ecosystem Proceedings." USDA Forest Service Gen. Tech. Rep. NE-23, pp. 137–158.

Gambel, J.N., and Likens, G.E., 1976. Calibration of collection procedures for the determination of precipitation chemistry. *Water, Air and Soil Pollution*, **6**: 241–258.

Gibb, J., R. Schuller, and R. Griffin, 1981. *Procedures for the collection of representative water quality data from monitoring wells.* Cooperative Ground Water Report No. 7, Illinois State Water Survey and Geological Survey. Champaign, Illinois.

Guy, H.P., and V.W. Norman, 1970. Field methods for measurement of fluvial sediment. In "Techniques of Water-Resources Investigations of the United States Geological Survey," Chapter C2, Book 3, U.S. Geological Survey, Washington, D.C.

Humenick, M., L. Turk, and M. Colchin, 1980. Methodology for monitoring ground water to uranium solution mines. *Ground Water* **18**: 262–275.

IHD-WHO Working Group on the Quality of Water, 1978. "Water Quality Surveys, Studies and Reports in Hydrology-23," 350 pp. United Nations Educational, Scientific and Cultural Organization, Paris, France, and World Health Organization, Geneva, Switzerland.

Karol, I.L., and L.T. Mijatch, 1972. Contribution to the planning of the station network for measuring the chemical composition of precipitation. *Tellus*, **24:5**, 421–427.

Kittrell, F.W., 1969. "A practical guide to water quality studies of streams." CWR-5, Federal Water Pollution Control Administration, U.S. Department of Interior, Washington, D.C. 135 pp.

Koehler, F.A., 1978. Simple sampler activation and recording system. *Journal of the Environmental Engineering Division, American Society of Civil Engineers*, **104**: 29–30.

Kunkle, S., W.S. Johnson, and M. Flora, 1978. "Monitoring stream water for land use impacts: a training manual for natural resources management specialists." Water Resources Division, National Park Service, Fort Collins, Colorado. 102 pp.

Likens, G.E., 1972. *The chemistry of precipitation in the Central Finger Lakes Region.* Tech. Rep. 50 Cornel Univ. Water Resources and Marine Sciences Center, Ithaca, New York. 47 pp.

Lodge, J.R., Jr., J.B. Pate, W. Basbergill, G.S. Swanson, K.C. Hill, E. Lorange, and A.L.

Lazrus, 1968. *Chemistry of United States Precipitation.* Final Report on the National Precipitation Sampling Network, National Center for Atmospheric Research, Boulder, Colorado. 66 pp.

Morrison, Robert D., 1983. *Groundwater Monitoring Technology.* Timco Manufacturing, Inc. 111 pp. Hastings, Pennsylvania.

Munger, J.W., and S.J. Eisenreick, 1983. Continental-scale variations in precipitation chemistry. *Environmental Science & Technology,* **17**: 32A–42A.

Nacht, S.J., 1983. Monitoring sampling protocol considerations. *Ground Water Monitoring Review.* Summer, 1983, pp. 23–29.

Nightingale, H., and W. Bianchi, 1979. Influence of well water quality variability on sampling decisions and monitoring. *Water Res. Bull.,* **15**, 1394–1407.

Pearson, F.J. Jr., and D.W. Fisher, 1971. Chemical composition of atmospheric precipitation in the northeastern United States. U.S. Geological Survey Water Supply Paper 1535P. Washington, D.C. 23 pp.

Pettyjohn, W., W. Dunlap, R. Cosby, and J. Keeley, 1981. Sampling ground water for organic contaminants. *Ground Water,* **19**: 180–189.

Scalf, M.R., J. McNabb, W. Dunlap, R. Cosby and J. Fryberger 1981. *Manual of Groundwater Sampling Procedures.* R.S. Kerr Environmental Research Lab. U.S.E.P.A., Ada, OK and National Water Well Assoc.

Schmidt, K., 1977. Water quality variations for pumping wells. *Ground Water,* **15**: 131–137.

Schuller, R., J. Gibb, and R. Griffin, 1981. Recommended sampling procedures for monitoring wells. *Ground Water Monitoring Review,* **1**: 42–46.

U.S. Department of Energy, 1985. "Procedures for the collection and preservation of groundwater and surface water samples and for the installation of monitoring wells." (2nd edition). GJ/TMC-08, UC-70A. Division of Remedial Action Projects, Technical Measurements Center, Grand Junction, Colorado 69 pp.

U.S. Department of Interior, 1977. "National Handbook of Recommended Methods for Water-Data Acquisition." U.S. Geological Survey, Office of Water-Data Coordination, Reston, Virginia. 990 pp.

U.S. Environmental Protection Agency, 1976. "Quality Criteria for Water. U.S. Government Printing Office, Washington, D.C. 256 pp.

U.S. Environmental Protection Agency, 1979. "Handbook for Analytical Quality Control in Water and Wastewater Laboratories." EPA-600/4–79–019. Environmental Monitoring and Support Laboratory, Cincinnati, Ohio.

U.S. Environmental Protection Agency, 1982. "Handbook for Sampling and Sample Preservation of Water and Wastewater." EPA-600/4–82–029. Environmental Monitoring and Support Laboratory, Cincinnati, Ohio. 402 pp.

U.S. Environmental Protection Agency, 1983a. "Technical support manual: Waterbody surveys and assessments for conducting use attainability analyses." Vol. I. Office of Water Regulations and Standards, Washington, D.C. 231 pp.

U.S. Environmental Protection Agency, 1983b. "Technical support manual: Waterbody surveys and assessments for conducting use attainability analyses." Vol. II: Estuarine systems. Office of Water Regulations and Standards, Washington, D.C. 186 pp.

U.S. Environmental Protection Agency, 1983c. "Water Quality Standards Handbook." Office of Water Regulations and Standards, Washington, D.C. 217 pp.

U.S. Environmental Protection Agency, 1986. *Quality criteria for water.* EPA-440/5–86-001. U.S. Government Printing Office, Washington, D.C.

Chapter 3

Water Quality Sampling Program Design

Federal Water Quality Legislation

Water quality is defined by use. Good water quality for irrigating corn is different from good water quality for mixing with whiskey. Water quality laws provide the broad goals and objectives for water quality monitoring. Specific goals or aims of water quality monitoring are agency-or problem-specific.

Any review of water quality legislation development would reveal the compromises in water quality legislative substance necessary for statute approval. Water quality laws only attempt to control or manage equitably the economic externalities associated with water quality degradation. The polarization between the "polluters" and the "regulators" obviously does not have much common ground. Often outside or political influences have more bearing on water quality legislation format. Because of this persistent dichotomy, water quality legislation is ambiguous. Introductory comments refer to the betterment of man and apple pie, those intangibles that can never be measured.

Water quality legislation is not as old as the water rights legislation, nor as confusing. A brief review of federal statutes is appropriate to see the legislative evolution from water-resources development to water-resources management. Legislation of most states parallels the federal legislation. Wildland hydrologists should be familiar with their own state legislation.

The first federal water quality legislation was enacted in 1886 to prevent deposit of refuse in New York Harbor. The busy port had ships from around the world and from other states. Often these ships would discharge refuse into New York Harbor. This flotsam drifted under the piers, and the concern was that the flotsam, particularly tar and oil products, could catch fire and destroy the piers. Such a disaster would interfere with commerce (the ability to do business) and that New York Harbor waters must be navigable to ensure commerce. Enforcement was given to the Department of the Army. Their presence in uniform and rifle was felt to be sufficient deterrent. This marks the beginning of the Army's involvement in water resources.

Since discharges from the ships were banned in New York Harbor, ships went to adjacent waters or tributaries and discharged there. The out of sight, out of mind philosophy is older than you think. In 1888, the 1886 act was extended to include adjacent and tributary waters of New York Harbor. Sewers and urban runoff was exempted, because there was no appreciation for public health and water quality.

In 1897, the Organic Act was passed to assure that forest resources were managed "to secure favourable conditions of water flows." This law has bearing on wildland management and minimum instream flows. The law was generally replaced in 1976 by the National Forest Management Act.

The Rivers and Harbors Act of 1899, commonly called the "Refuse Act," was in effect until 1972 with the passage of PL92-500. This act had the Department of the Army regulate all discharges (point-source) in navigable waters. The definitions of navigable waters have varied over time; however, if the water had some role in commerce, it was generally considered navigable. Smaller streams that floated logs to the mills were thus "navigable" and expanded the regulation area of the Army. Liquid wastes from sewers were again excluded from this law.

Recognition of human diseases with water pollution was made official in 1912. The United States Public Health Service could investigate, but not correct, such water quality problems.

The Oil Pollution Act was passed in 1924. This law controlled oil discharges in coastal waters to minimize beach and shellfish impacts and to reduce fire hazards around the seaports.

A water-pollution control act was passed through both houses, but was vetoed by President Roosevelt in 1938. Previous attempts to pass a federal water-pollution act had been made, but a compromise act was never reached by both houses. Roosevelt vetoed the 1938 act because of fear that it would interfere with commerce in the United States. Nothing should threaten the economic growth of the United States.

A threshold of action occurred in 1948 with the passage of the first Federal Water Pollution Control Act (FWPCA) (PL80-845). The law was set up for 5 years, with primary control of water pollution given to the individual states. The U.S. Public Health Service was again available for consultation but not problem correction. Federal funds were available for water-pollution research and assistance. The 1948 act was extended by the FWPCA Amendment of 1955 (PL82-579).

The first permanent federal water-pollution legislation was passed as the 1956 FWPCA Amendments (PL84-660). This act established the Federal Water Pollution Control Administration under the U.S. Public Health Service. A national network for water quality surveillance and the associated database was established. This network is run by the U.S. Department of Interior, Geological Survey. The act strengthened research and training aspects of the federal grants-in-aid to states and funded public waste treatment plant construction and upgrades. Up to this point, there had only been one criminal–civil suit on water quality, which was dismissed because of the defendant's inability to prove plaintiff responsibility.

The 1961 FWPCA Amendments (PL87-88) transferred the federal administration of water-pollution control from the U.S. Public Health Service to the Department of Health, Education, and Welfare. Their jurisdiction covered navigable interstate waters to eliminate double standards for interstate and intrastate waters.

The Water Quality Act of 1965 (PL89-234) marked a significant change in the federal legislation attitude. The water "pollution" titles were changing to water "quality." Water "quality" implied a more positive and preventive attitude rather than the negative and corrective attitude of water pollution. This act required state standards to be used by the federal government for interstate waters. Enforcement was through enforcement conferences. States designed standards for receiving waters of discharges to meet stream standard. Each discharge (source) had to be individually evaluated for its impact on the receiving stream. The enforcement conference effort was valiant, but numerous political, technical, and legal weaknesses existed. There was no comment on non-point-source pollution.

The FWPC Administration was transferred to the Department of Interior by the 1966 Clean Water Restoration Act (PL39-753). The only other change by this law was the federal funding formula for state wastewater treatment plant construction and upgrade.

The Water Quality Improvement Act (PL91-224) increased federal authority over oil spills and hazardous wastes. The act replaced the 1924 Oil Pollution Act. Of most impact, this is the first federal attempt at developing a liability principle for water quality degradation.

The National Environmental Protection Act (NEPA) of 1969 was signed by President Nixon in his pajamas on January 1, 1970. NEPA combines a federal policy with environmental protection. An environmental impact statement must be prepared if the proposed activity is to have significant environmental impact, is federally funded, or is on federal lands. Most states have State Environmental Protection Acts (SEPAs) that, in general, are stronger than NEPA.

Creative Order Reorganization Plan Number 3 was passed in 1970 to create the U.S. Environmental Protection Agency. This single agency was to oversee regulation and enforcement on air quality, water quality, and solid and hazardous wastes.

The 1972 FWPCA Amendments (PL92-500) had the most impact on wildland water quality because non-point-source pollution was recognized. The law stated that the nation would have zero discharge by 1985, with all waters swimable and fishable. The legislative process required to pass this law included compromises, lobbying, late night sessions, and a override of Nixon's veto. Briefly, the wildland hydrologist should be familiar with the following sections:

Section 208, Non-Point-Sources. Federal funds were available for states to identify existing non-point-sources of water quality pollution. The non-point-sources were to be reduced or eliminated by development of best management practices. These practices, when implemented, were designed to meet receiving water standards.

Section 404, Army Corps of Engineers Permit. If any land-use activity involved moving more than 50 cubic yards of material, affected more than 10 acres of water in a lake or pond, or streamflow was greater than 5 cubic feet per second, a 404 permit application would be required from the Army Corps of Engineers. The permit application includes reviews by both state and federal fisheries agencies.

The first attention to groundwater quality was in the 1974 Safe Drinking Water Act. The law established minimum national standards for public water supplies and mandated the Environmental Protection Agency to regulate state programs to control underground injection wells for waste disposal of harmful wastes. Fluids for oil and gas extraction were exempt.

The Clean Water Act of 1977 (FWPCA Amendments of 1977, PL92-217) were refinements to PL92-500. This law strengthened and clarified the state roles in water-resource classification and standard development.

The Resource Conservation and Recovery Act (RCRA) of 1976 and the Comprehensive Environmental Response, Compensation and Liability Act of 1980 (CERCLA) address the issue of hazardous wastes. The laws are continually amended, and if the wildland hydrologist is sampling water quality in a RCRA or CERCLA situation, familiarity with the latest law is essential.

The Federal Water Pollution Control Act Amendments of 1972 (PL92-500) established a legislative framework from which an assessment of the sources and extent of non-point-source pollution and development of guidelines and procedures to control non-point-source pollution could be made. Non-point-source water pollution can be controlled only by managing the type of activity that takes place on a watershed. Many non-point-sources of pollution are present naturally in varying quantities. It must be assumed that Congress was not intending that water in any given stream be of a quality suitable for a specific downstream use; rather, Congress was aiming to ensure that streams maintain or progress toward the ecosystem that existed in their pristine state. Present legislation allows for annual and longer-term fluctuations and for the probability of major

Water quality from undisturbed forested watersheds is usually good. Land-use activities done with the incorporation of best management practices (BMPs) help ensure water-quality maintenance.

disturbance by natural events (Meier, 1976). Presently no single index for water quality has found general acceptance because of lack of consensus as to the definition of water quality and to the interpretation of index values. Also, there is no method for assessing and comparing the performance of indices (Stednick, 1980).

Land-use activities and the resultant water quality changes, either positive or negative, are considered non-point-source pollution. Non-point-source pollution was previously addressed in Section 208 of the 1972 Amendments. The 208 program was part of waste-treatment management plans to be implemented by individual states.

The most effective unit of available prevention and control of non-point-source pollution constitutes what is known as a "best management practice" (BMP). This is an administrative creation and initially consists of a combination of practices that are determined after problem assessment. Practices are examined, with appropriate public participation, for practicality and effectiveness in preventing or reducing the amount of pollution generated by diffuse sources, to a level compatible with water quality goals (McClimans et al., 1979; Meier, 1975; Rice et al., 1975).

An alternative to managing non-point-source pollution by using water-quality standards is through standardization of land-use activities. However, one must realize that the physical characteristics of the land will best define which techniques should be applied. The flexibility that is provided administrators in the development of BMPs is recognition of the difficulties associated with controlling pollution generated from non-point-sources.

The Water Quality Act of 1987 (PL100-4) has the potential to identify technology to better define the impact of land use activities on water quality. The Water Quality Act of 1987, the Clean Water Act of 1977 (PL95-217), and the Federal Water Pollution Control Act Amendments of 1972 are collectively referred to as the Clean Water Act.

Specifically, Subsection 316, Non-Point-Source Pollution Management requires that each state prepare, within 18 months of enactment, an assessment report and a management program for non-point-source pollution. The assessment report will identify (1) those waters that are unlikely to comply with water quality standards without additional controls on non-point-sources of pollution and (2) the non-point-sources causing the problem. The management program, which covers a 4-year period, will include (1) identification of measures to control the non-point-source pollution identified in the assessment report, (2) identification of programs to implement those measures, (3) certification that the state laws have adequate authority to implement the program, (4) identification of sources of all funding for non-point-source pollution control, and (5) a schedule for expeditious implementation of the program.

Assessment of potential water quality changes due to land-use activities may be measured by water quality standard compliance and/or definition of cumulative watershed effect. Cumulative impact is the impact that results from the incremental impact of the action when added to other past, present, and reasonably foreseeable future actions regardless of what agency or person undertakes such other actions. Such impacts can result from individually minor but collectively significant actions taking place over a period of time (40CFR1508.7). The concept of cumulative impacts is especially important in regard to the effect of deposited sediment on fish habitats. Most cumulative impact assessment efforts to date use physical characteristics of the stream channel.

Cumulative impacts to a stream transcend ownership boundaries. This is of particular concern in mixed-ownership watersheds (checkerboard lands) where no single agency or owner has overall regulatory authority. The federal regulatory authorities have approached cumulative impacts from two different directions to date. The first approach is an attempt to develop alternative regulatory perspectives to better define cumulative watershed effects. The second is to determine an allowable impact akin to performance standards. However, the water quality standards give a simple positive–negative answer: yes, water quality standards have been exceeded; or no, water quality standards have not been exceeded. There is a need to better quantify or evaluate water quality concentrations of various parameters that are above background but below the allowable concentration or standard. The alternative regulatory perspective is institutional and can be evaluated only after implementation.

Water quality is an expression of all hydrologic processes occurring in a watershed. Thus, land-use activities and their potential environmental impact may be assessed by water quality changes. Perhaps the easiest to measure, and certainly the most documented, are the effects of land use on water yield.

Water Quality Standards And Criteria

The development of water quality legislation has been the result of compromises over time. Unfortunately, federal and state water quality laws are not usually based on sound environmental or economic judgment. Water quality standards are established by legislative authority and subject to lobbying influence and may not have a regulatory effect on certain receiving waters or stream classifications. Standards are state-specific and developed after inventory sampling of state waters. The standards are relatively easy to measure, and usually any departure is visible; for instance, pH or temperature change may kill biota, and extremely turbid waters are noticeable (Table 3.1). The public is still the best environmental watchdog.

Water quality criteria are not established by legislative authority and not subject to the above problems. Often criteria provide limits

TABLE 3.1.
Water Quality Standards for General Classifications of Colorado Water

Parameter	Recreation	Aquatic	Agriculture	Domestic
Dissolved oxygen (mg L^{-1})	Aerobic	6.0	Aerobic	Aerobic
pH (standard units)	6.5–9.0	6.5–9.0		5.0–9.0
Turbidity (NTU)^a				1.0
Temperature (°C)		Maximum 20°C or 3°C increase		
Fecal coliform (number per 100 mL)	200–20000			0
Ammonia (mg L^{-1} as N)		0.02		0.5
Nitrate (mg L^{-1} as N)			100	
Sulfate (mg L^{-1})				250
Chloride (mg L^{-1})				250

^{a} NTU = Nephelometric Turbidity Unit.

for water quality constituents not addressed by the standards. The "E.P.A. Red Book" or Quality Criteria for Water published in 1976 by the U.S. Environmental Protection Agency was updated as the Gold Book in 1986 and is considered the reference for water quality criteria. Criteria do not have direct regulatory use, but they form a basis for judgment.

Variability in water quality exists in streams and between streams; thus a single numerical value may have limited applicability. Certain organisms may become adapted to existing water quality that may be considered extreme (or lethal) in other areas. Organism criteria (fresh or marine) are based on bioassay or tank studies and are based on an extensive review of species specific studies. The criteria may include potable limits and/or fresh and/or marine organism limits. Potable limits have been developed with the U.S. Public Health.

Water Quality Monitoring Programs

Water quality monitoring may be divided into four types of programs: (1) cause and effect, (2) compliance, (3) baseline, and (4) inventory (Ponce, 1980).

Stream discharge measurements are an important component of water quality monitoring. If the water quality station is long-term or has high value, artificial control sections are warranted. Here a broad-crested compound weir is used to measure the stream stage. Flynn Creek, Oregon Coast Range.

Cause-and-effect sampling is to measure treatment or land management activity changes on water quality. This sampling is usually paired sampling on a random or systematic sampling frequency. Sampling is short-term but with high intensity. This sampling program is used for treatment versus control or pre- versus post-treatment studies.

Compliance sampling is to ensure that state and/or federal water quality standards are being met. The compliance sampling program is used by regulatory agencies. Sampling duration is intermittent and the intensity variable.

Baseline sampling is long-term sampling of variable intensity used for trend analysis. Federal agencies would be involved in sampling of this nature, since few state agencies have the budget. Land use may be changing in this sampling program.

Inventory sampling is to determine what the water quality is in a specific area. The sampling duration is usually long and the inten-

Short-term water quality sampling stations may be measured for stream discharge by the sag-tape method using permanent benchmarks on either side of the stream. Cache la Poudre River, Colorado.

sity low, but depends on definition of the area's water quality. The U.S. Geological Survey will inventory sample for no less than 10 normal years. This time period is felt to include at least one high and one low streamflow year of a 10 percent recurrence interval.

Water quality as a subject continues to increase in importance. Concomitantly, the literature on water quality is increasing; however, attention to design of water quality monitoring programs has only recently come about. There are several approaches to the design of water quality monitoring programs. Presented here is a synthesis of ideas. Not all steps in this format are necessarily followed, as it depends on the water quality monitoring purpose. Nonetheless, programs may be developed by following these steps:

1. Determine study objectives and importance of each.
2. Assign task responsibilities.
3. Review literature on subject material.
4. Express objectives in statistical terms.

5. Establish data analysis techniques and interpretation.
6. Allocate budget on a per objective basis.
7. Characterize the monitoring area.
8. Determine water quality variables to be monitored.
9. Do presampling and know lab availability.
10. Determine sampling station locations.
11. Determine sampling frequency and sample type.
12. Develop operating plan.
13. Develop data management system.
14. Develop feedback mechanisms.
15. Prepare final report with network design and results.

STUDY OBJECTIVES

Specific study objectives need to be formulated before sampling begins. The objectives should be expressed as a hypothesis that may be statistically tested. Objectives should be prioritized or ranked, especially if under time or budget constraints.

TASK RESPONSIBILITIES

As obvious as it may seem, it is important to define and assign specific task responsibilities. This is true for not only the individuals in your study group but also between agencies and individuals in your professional career.

LITERATURE REVIEW

Currently, over 600 pieces of information are published per day in the United States! Thus, chances of finding information on a water-quality program of any choosing are quite high. A literature review will offer answers to many questions.

Sampling techniques, water quality variables, and data characteristics are just a few of the many questions to be answered. Previous and similar studies will assist in hypothesis formulation. However, be cautious in extrapolating data from a previous study to your study area.

STATISTICS

Each objective should be stated in statistical terms. The larger studies will have several objectives, and for convenience you may enumerate them. Most null hypotheses are that the means are equal (treatment vs. control), unless data indicate otherwise. Be careful of a hypothesis stating that A is greater than or equal to B. These one-tailed tests may show a bias.

ANALYSIS TECHNIQUES

Preliminary sampling is advantageous for both lab technique review and assessment of the sampling location. You have to be comfortable with the lab analysis to recognize potential errors, either random or systematic. This familiarity is helpful when doing the statistics. A measured value of zero or below detection limit is still data and should be treated as such. It is very rare to justify "throwing out" data points. Remember significant figures are only as good as your lowest value. Thus, if you measure pH to the nearest 0.02 unit, the average (or other expression) is to the nearest 0.02 unit.

WATER QUALITY VARIABLE SELECTION

The selection of water quality variables for a water quality sampling program will often depend on the purpose of the study, legislative mandate, the sampler's interest, or the available budget. Don't be afraid to analyze for parameters that aren't popular; if they can add information to the database, they can be sampled. The relation of a particular constituent to the beneficial use of the water may be used as the first step in variable selection. Literature review is the best way to help select water quality parameters for any monitoring program.

SAMPLING STATION LOCATION

Sampling station location will depend on the study objectives. Sampling stations can be a single site for compliance or inventory sampling, a pair of sites for cause-and-effect monitoring, or several stations constituting a water quality monitoring network. For treatment comparison, homogeneous units need to be delineated. Homogeneous units are where all physical variables are the same, except the treatment or activity. Thus, the difference in water quality may be attributed to the defined activity.

Sampling stations need to be accessible for all flow conditions that will be sampled. If on private property, written permission carried with you will help explain the situation to the sheriff. Stream discharge needs to be taken with every sample, thus a stream section with some form of control is helpful.

Avoid sampling near a sediment plume or point bar or other channel feature that is atypical of the reach. If sampling below a tributary on the higher-order stream, make sure that the streams are well mixed. Mixing in both the vertical and horizontal directions will often occur in five to seven times the channel width. Point-source pollution should be recognized when locating a sampling station.

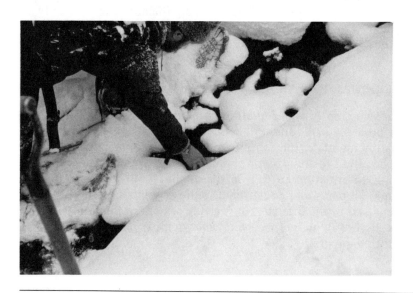

Stream water quality sampling during the winter presents different logistical problems. Site access should be considered during the sampling program design. Note snowshoes. Hourglass Creek, Colorado.

SAMPLING FREQUENCY

Sampling frequency is a function of the study objectives and will thus reflect any budget constraints. The minimum sample size may be calculated (see Chapter 20) for a given error and confidence term. However, if extremes or regressions or probabilities are being tested, more sampling may increase the confidence level. The tradeoff is between budget and time allowed.

Stream discharge measurements should be made each time a water-quality sample is taken. Often relationships exist between constituent concentration and stream discharge. Sampling frequency calculations may be compounded by streamflow–constituent concentration relations. Normal unbiased sampling is best when there is no relation between streamflow and constituent concentration; a graph of these two variables quickly illustrates any relation. The relations may be flow-driven, where constituent concentration increases with discharge. Such a relation may include total suspended solids, orthophosphate phosphorus, or fecal coliform. The opposite relation is flow–dilution when constituent concentrations decrease with increased streamflow. Most conservative constituents such as nitrate or potassium have this relation. Finally, the flow–dilution-driven relation is when concentrations decrease as flow increases, and then begin to increase as the flow continues to increase. The latter increase is when different flow routing mechanisms are active. Often alkalinity or conductivity have such a flow relation.

Consideration of these relations should be made when determining sampling frequency for a given site with its annual hydrography, frequency or high flows or low flows, and the specific land-use activity timing (if any). These relations are important when assessing water quality for low or high concentrations, as they relate to standards and criteria or another critical period, and when determining the chemical or nutrient flux, where volume-weighted concentrations are required. The ability to predict constituent concentration by streamflow has a variety of applications in water-quality monitoring.

This is a capsule summary for designing a water-quality monitoring program. Each sampling program may be different, and professional judgment will guide you in designing a program to answer the study objectives within the time and budget constraints.

Suggested Readings

Averett, R.C., 1978. *Selecting water quality parameters for the specified program or study goals*. U.S. Geological Survey, Water Resources Division, Denver, Colorado. Unpublished manuscript. 17 pp.

Environment Canada, 1983. *Sampling for water quality*. Water Quality Branch, Inland Waters Directorate, Ottawa, Canada. 55 pp.

Hohenstein, W.G., 1987. Forestry and the Water Quality Act of 1987. Journal of Forestry **85**: 5–8.

McClimans, J.R., J.T. Gebhardt, and S.P. Roy, 1979. *Perspectives for "silvicultural best management practices"*. Applied Forestry Research Institute. State University of New York. Research Report No. 42. 113 pp.

McKee, J.E., and H.W. Wolf (eds.), 1963. "Water quality criteria." Publication No. 3-A. California State Water Resources Control Board. Sacramento, California 548 pp.

McNeeley, R.N., V.P. Neimanis, and L. Dwyer, 1984. "Water quality sourcebook: a guide to water quality parameters." Environment Canada, Inland Waters Directorate, Water Quality Branch, Ottawa, Canada. 88 pp.

Meier, M.C., 1976. Research needs in erosion and sediment control. *In* "Soil Erosion: Prediction and Control." Soil Conser. Society of America. Special Publ. No. 21. pp. 86–89.

Perry, J.A., R.C. Ward, and J.C. Loftis, 1984. Survey of state water quality monitoring programs. *Environmental Management* **8(1)**: 21–26.

Phillips, J.D., 1988. Non-point source pollution and spatial aspects of risk assessment. *Annals of the Assoc. of Amer. Geog.* **78(4)**: 611–622.

Ponce, S.L., 1980. *Water quality monitoring programs*. WSDG Technical Paper WSDG-TP-00002. U.S. Department of Agriculture, Forest Service, Watershed Systems Development Group, Fort Collins, Colorado. 68 pp.

Potyondy, J.P., 1980. *Technical guide for preparing water quality monitoring plans*. U.S. Department of Agriculture, Forest Service, Intermountain Region, Ogden, Utah. 74 pp.

Resh, V.H., and D.G. Price, 1984. Sequential sampling: A cost-effective approach for monitoring benthic macroinvertebrates in environmental impact assessments. *Environmental Management* **8(1)**: 75–80.

Rice, R., R. Thomas, and G.W. Brown, 1975. *Sampling water quality to determine the impact of land use on small streams*. Unpublished paper. ASCE Watershed Report Meeting, Logan, Utah.

Sanders, T.G., R.C. Ward, J.C. Loftis, T.D. Steele, D.D. Adrian, and V. Yevjevich, 1983. "Design of networks for monitoring water quality." Water Resources Publications, Littleton, Colorado. 328 pp.

Sherwani, J.K., and D.H. Moreau, 1975. "Strategies for water quality monitoring." Water Resources Research Institute of the University of North Carolina, Raleigh, North Carolina. 137 pp.

State of California, 1990. *Forest practices cumulative impact assessment process*. Tech. Role Addendum No. 2. State Board of Forestry, Sacramento, California. 25 pp. In press.

Stednick, J.D., 1980. Alaska water quality standards and BMPs. *In* "Proceedings of Watershed Management Symposium." Amer. Society of Civil Engineers, Boise, Idaho. pp. 721–730.

Steele, T.D., 1985. Strategies for water quality monitoring, pp. 311–344. *In* J.C. Rodda (ed.), "Facets of Hydrology," Volume II. John Wiley and Sons Ltd., New York.

U.S. Environmental Protection Agency, 1976. "Quality criteria for water." U.S. Government Printing Office, Washington, D.C. 256 pp.

U.S. Environmental Protection Agency, 1982. "Handbook for sampling and sample preservation of water and wastewater." EPA-600/4-82-029. Environmental Monitoring and Support Laboratory, Cincinnati, Ohio. 402 pp.

U.S. Environmental Protection Agency, 1983a. "Water quality standards handbook." Office of Water Regulations and Standards, Washington, D.C. 217 pp.

U.S.Environmental Protection Agency, 1983b. "Technical support manual: Waterbody surveys and assessments for conducting use attainability analyses." Vol. II: Estuarine Systems. Office of Water Regulations and Standards, Washington, D.C. 186 pp.

U.S. Environmental Protection Agency, 1983c. "Water quality standards handbook." Office of Water Regulations and Standards, Washington, D.C. 217 pp.

U.S. Environmental Protection Agency, 1986. *Quality criteria for water.* EPA-440/5-86-001. U.S. Government Printing Office, Washington, D.C.

Chapter 4

pH

Purpose: To learn the methodology and applications of pH measurements using the standard glass electrode method.

The pH of water is a determining factor in almost every natural (and treatment) process, a critical component of biologic systems, and an important tool in measuring other water quality parameters such as alkalinity. The pH parameter is a measure of the hydrogen ion activity, which in dilute solutions may be considered approximately equivalent to the hydrogen ion concentration. The symbol "p" represents the negative logarithm. Pure water is very slightly ionized and at equilibrum:

$$[H^+] \, [OH^-] = K_w = 1.01 \times 10^{-14} \text{ at } 25°C$$

$$[H^+] = [OH^-] = 1.005 \times 10^{-7}$$

where $[H^+]$, $[OH^-]$ indicate activities of these ions in moles per liter and pH $= -\log_{10} [H^+]$. Thus pH of pure water $= 7.001$, but pH may range from 0 to 14.

The basic principle of electrometric pH measurement is determination of the activity of the hydrogen ions by potentiometric measurement using a standard hydrogen electrode and a reference electrode. The electromotive force (emf) produced in the glass electrode varies linearly with pH. The glass electrode, when used in combination with a calomel reference electrode, produces a charge of 59.1 mV per pH unit at 25°C. The pH measuring instrument is calibrated potentiometrically using standard buffer solutions. Sample pH is determined by interpolation or extrapolation.

In the measurement of pH values of water samples, the electrodes must be thoroughly rinsed with buffer solutions between samples and after calibrating. The electrodes should be kept free of oil and grease and stored in a buffer solution when not in use. In testing samples containing gaseous or volatile components that affect the pH value, any handling technique such as stirring or temperature change may cause loss of such components and thereby introduce error.

Procedure

BUFFER SOLUTIONS

The pH meter must be calibrated against solutions of known pH. Buffer solutions are commercially available as solutions or powders that are diluted to volume. Common buffer solutions are for pH values of 4.0, 7.0, and 10.0. Buffer solutions are available for other pH values, but the added product expense seldom is justifiable.

The pH meter should be calibrated with two buffer solutions that most approximate the sample pH. In most surface waters, the pH ranges from 6 to 8, so buffers of 4 and 7 or 7 and 10 must be used. The buffer solution temperature should be recorded and the appropriate temperature compensation made on the pH meter. Remember to use some of the stock buffer solution in a small labeled beaker for calibration. The used buffer solution should be discarded after use, not returned to the stock bottle, to avoid buffer contamination.

If the instrument does not calibrate for both buffers, that is, the instrument reads 4.0 for the 4.0 buffer and 6.9 for the 7.0 buffer, split the error between the buffers to ensure readings of 4.05 and 6.95. If the calibration error is more significant, the instrument needs servicing. If the instrument is portable, check the batteries. Finally, the buffer solutions may be contaminated, or simply have deteriorated over time.

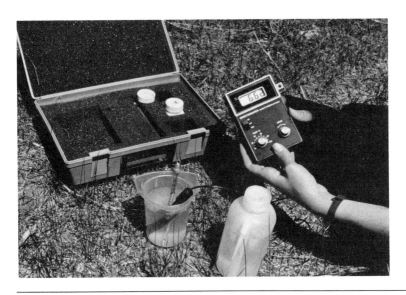

Portable instruments such as this pH meter may be used for in-field analysis. The sample used for these analyses should be discarded and not returned to the collected sample.

PROCEDURES

Repeat the measurement procedure at least three times for each sample. Keep electrodes immersed in buffer solution. Before use, remove the electrodes from the buffer solution and rinse with distilled water over a "wash" beaker. Dry electrodes gently with soft tissue. Standardize the instrument with buffer solution. Note the pH and temperature of the buffer. Remove electrodes, rinse thoroughly, and dry.

If the sample temperature is different than the buffer temperature, let the electrodes equilibrate with the sample. Measure the sample temperature and set the temperature compensator on the pH meter to the measured temperature. Note and record pH and temperature. Rinse electrodes and immerse in buffer until the next measurement.

TYPICAL VALUES	pH (Standard Units)
Cache la Poudre River at canyon mouth near Fort Collins, Colorado USGS Station 0675200	7.2
Tualatin River at West Linn, Oregon USGS Station 14207500	7.4
Rogue River near Agness, Oregon USGS Station 14372300	7.5
Colorado River at Stateline, Colorado USGS Station 09163500	7.9
Colorado River at Imperial, California USGS Station 09429490	7.9
Skagit River at Marblemount, Washington USGS Station 12181000	7.3
Yellowstone River near Livingstone, Montana USGS Station 06192500	7.9
Arkansas River near Coolidge, Kansas USGS Station 07137500	8.0
Cumberland River near Grand River, Kentucky USGS Station 03438220	7.6
Ogeechee River near Eden, Georgia USGS Station 02202500	6.9
North Fork Whitewater River near Elba, Minnesota USGS Station 05376000	8.3

Review

KEYWORDS

pH
hydrogen ion activity
glass electrode
reference electrode
dissociation

STUDY QUESTIONS

1. Calculate the mean and standard deviation for the pH of each sample. Give three possible reasons for the variability in your data.

2. We know that, at equilibrium

$$K_W = [H^+] \, [OH^-] + 1.01 \times 10^{-14} \quad (25°C)$$

For pure water, with no other ions present, we can say:

$$[H^+] = [OH^-] = 1.005 \times 10^{-7} \quad \text{at equilibrium, } 25°C$$

But when other ions are present, this is not true. For instance, 0.01 N solution of HCl (in distilled water) is subject to the following equilibria relationships:

$$[H^+] \, [OH^-] = K_w = 1.01 \times 10^{-14} \quad \text{at } 25°C$$

$$[H^+] \, [Cl^-] = K_a = 10^{-3} \quad \text{at } 25°C$$

The stoichiometric reactions involved are

$$H_2O \rightleftharpoons H^+ + OH^-$$

$$HCl \rightleftharpoons H^+ + Cl^-$$

If $[Cl^-] = 0.01$ mol L^{-1}, what is the equilibrium pH, and the equilibrium $[OH^-]$ of this solution at 25°C? Assume that ion activities are equal to their concentrations.

3. A dilute HCl solution contains 50 mg L^{-1} of $[Cl^-]$. What is the pH of the solution? (Remember that equilibrium constants use concentrations in moles per liter).

(continued)

Review (Continued)

4. Acid rain is caused, in part, by SO_2 emissions that are hydrolyzed and oxidized to SO_4^{2-} in the atmosphere. Discuss briefly and in general terms, how the presence of SO_4^{2-} in precipitation decreases the pH of the precipitation.

SAMPLE DATA SHEET

Analysis: pH   Instrument name:

Analyst: Instrument serial number:

Date: Calculations done by:

Start time: Calculations checked by:

  End: Comments:

Procedure:

SAMPLE IDENTIFICATION:	REPLICATE	MEASUREMENT
Standard	1	
	2	
Reported Value:	3	
	Mean:	
	Standard Deviation:	
	In Control?	

SAMPLE IDENTIFICATION:	REPLICATE	MEASUREMENT
	1	
	2	
Reported Value:	3	
	Mean:	
	Standard Deviation:	
	In Control?	

SAMPLE IDENTIFICATION:	REPLICATE	MEASUREMENT
	1	
	2	
Reported Value:	3	
	Mean:	
	Standard Deviation:	
	In Control?	

Suggested Readings

American Public Health Association, 1985. "Standard Methods for the Examination of Water and Wastewaters." Published by the Amer. Public Health Assoc., Amer. Water Works Assoc., and Water Poll. Control Fed., 16th Edition. pp. 429–437.

Cogbill, C.V., and G.E. Likens, 1974. Acid precipitation in the northeastern United States. *Water Resources Research* **10**: 1133–1137.

Grant, L., 1972. On the relation between pH and the chemical composition in atmospheric precipitation. *Tellus*, **24**: 550–560.

Kunkle, S., and J. Wilson, 1984. *Specific conductance and pH measurements in surface waters: An Introduction for park natural resource specialists.* WRFSL Report No. 84–3. National Park Service, Water Resources Field Support Laboratory, Colorado State University, Fort Collins, Colorado. 51pp.

Chapter 5

Conductivity

Purpose: To illustrate the use of conductivity measurements in water analysis, and to illustrate the use of conductivity measurements to approximate total dissolved solids.

Conductivity is a convenient, rapid method of estimating the amount of dissolved solids. It is a numerical expression of the ability of an aqueous solution to convey an electric current. This property is related to the total concentration of ionized substances and their respective concentrations, mobility, and valence, and to the temperature at which the measurement is made. Solutions of most inorganic acids and bases are relatively good conductors. Organic compounds that do not dissociate in aqueous solutions are not good conductors.

The physical measurement of conductivity is usually in terms of resistance. The resistance of a conductor is inversely proportional to its cross-sectional area and directly proportional to its length. Thus, for meaningful results, the conductivity cell characteristics must be known.

For a given cell with fixed electrodes, the ratio of the distance d between the electrodes to their area A, d/A, is constant and is defined for a given temperature as the cell constant, q:

$$k = k_m q$$

where k_m is the measured value of conductivity for specific conditions, k is the specific conductance, and q is a cell constant. This
conversion standardizes measured conductivity (in mhos) to units
that are comparable with data from other conductivity meters (mhos
per centimeter).

Conductivity measurement also is used to indicate the concentration of total dissolved solids in natural waters (see Figure 5.1). An
empirical relation between dissolved salt concentration and conductivity has been suggested. Total dissolved solids (mg L^{-1}) is equal to
the conductivity multiplied by a factor between 0.55 and 0.9 depending on the soluble constituents in a specific water.

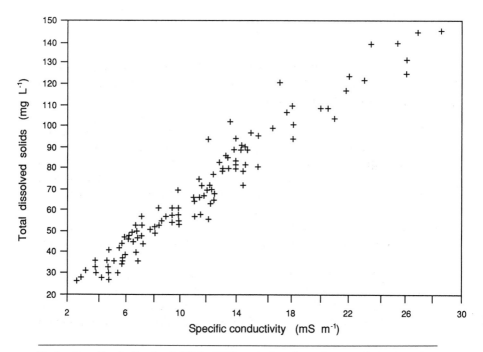

Figure 5.1. Total dissolved solids (TDS mg L^{-1}) is linearly correlated to specific
conductivity (mS m^{-1}). Such relations are site-specific.

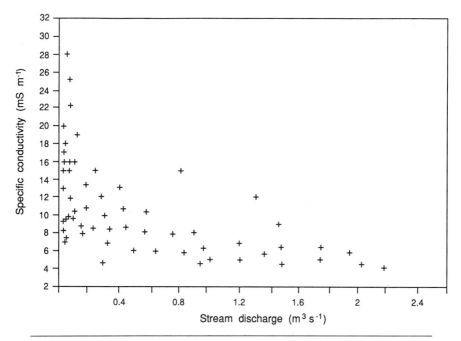

Figure 5.2. A flow–dilution relationship between specific conductivity and stream discharge. As stream discharge increases the conductivity decreases.

Conductivity is customarily reported as micromhos per centimeter ($\mu \mho \, cm^{-1}$). In the International System of Units (SI) the reciprocal of the ohm (Ω) is the Siemen, and conductivity is reported as milliSiemens per meter (mS m^{-1}; see Figure 5.2); 1 mS m^{-1} = 10 $\mu \mho \, cm^{-1}$.

Procedure

Most conductivity meters now are self-contained and allow for temperature compensation to 25°C if the solution is not at 25°C. The function switch is set for temperature and the solution temperature recorded. The instrument (see owner's manual) may or may not have automatic temperature compensation.

Place the conductivity cell in the unknown solution. The electrodes must be submerged, and the electrode chamber must not contain trapped air. Tap the cell to remove any bubbles and dip it two or three times to assure proper wetting. If the instrument is direct reading, record the conductivity. Older conductivity meters require the reading to be multiplied by the cell constant; again refer to the owner's manual.

TYPICAL VALUES	Conductivity (μS cm^{-1} at 25°C)
Cache la Poudre River at canyon mouth near Ft. Collins, Colorado	104
Tualatin River at West Linn, Oregon	142
Rogue River near Agness, Oregon	100
Colorado River at Stateline, Colorado	880
Colorado River at Imperial, California	1220
Skagit River at Marblemount, Washington	53.1
Yellowstone River near Livingstone, Montana	219
Arkansas River near Coolidge, Kansas	3360
Cumberland River near Grand River, Kentucky	178
Ogeechee River near Eden, Georgia	54
North Fork Whitewater River near Elba, Minnesota	522

Review

KEYWORDS

conductivity
specific conductance
resistance
conductivity meter

STUDY QUESTIONS

1. Speculate on the effects of turbidity on solution electrical conductivity.
2. How can solution electrical conductivity be used as a quality-control measurement?
3. Why should electrical conductivity be measured in the field?
4. Which solution has the higher electrical conductivity:
   (a) 1 M HCl   or   1 M H_2SO_4?
   (b) 1 M KCl   or   2.0 M HCl?
   (c) 1 M KCl   or   0.5 M $CaCl_2$?

SAMPLE DATA SHEET

Analysis: Conductivity Instrument name:

Analyst: Instrument serial number:

Date: Calculations done by:

Start time: Calculations checked by:

 End: Comments:

Procedure:

SAMPLE IDENTIFICATION:	REPLICATE	MEASUREMENT
Standard	1	
	2	
Reported Value:	3	
	Mean:	
	Standard Deviation:	
	In Control?	

SAMPLE IDENTIFICATION:	REPLICATE	MEASUREMENT
	1	
	2	
Reported Value:	3	
	Mean:	
	Standard Deviation:	
	In Control?	

SAMPLE IDENTIFICATION:	REPLICATE	MEASUREMENT
	1	
	2	
Reported Value:	3	
	Mean:	
	Standard Deviation:	
	In Control?	

Suggested Readings

American Public Health Association, 1985. "Standard Methods for the Examination of Water and Wastewaters." Published by the Amer. Public Health Assoc., Amer. Water Works Assoc., and Water Poll. Control Fed., 16th Edition. pp. 76–80.

Kunkle, S.H., 1984. *Specific conductance and pH measurements in surface waters: an introduction for park natural resource specialist*. WRFSL Report No. 84–3. National Park Service, Water Resources Field Support Laboratory, Colorado State University, Fort Collins, Colorado. 51 pp.

Kunkle, S.H., 1972. Effects of road salt on a Vermont stream. *J. Amer. Water Works Assoc.* **64(5)**: 290–295.

Chapter 6

Water Quality Sampling Field Methods

Water quality data are only as good as the sample. This chapter synthesizes the techniques that should be done in the field for surface-water quality sampling. The most important field measurement is stream discharge (or stream stage) (review Chapter 2; see also Figure 6.1). Physical parameters of stream (or other wildland waterbody) include temperature, pH, conductivity, dissolved oxygen, and alkalinity.

Stream Discharge

Equipment List for Stream Discharge

1. Flow meters—AA Price or Pygmy with wading rod, headphone, extra batteries, and copy of rating table (or direct reading instrument).
2. Tape measure—cloth tape is recommended.
3. Stakes—end stakes for anchoring tape measure to stream banks.

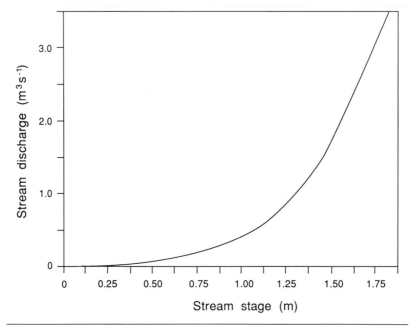

Figure 6.1. A stage–discharge curve. The stream gauging station records stream stage, which is converted to stream discharge.

4. Stopwatch.
5. Waders—hip or chest.
6. Pencils, markers, and labels.
7. Discharge and sampling forms.
8. Calculator.

Gaging Site Selection

1. Purpose—have reasons for sampling particular spots.
2. Access—near a road or trail.
3. Locatable—a distinctive feature nearby will help locate sampling point on the map.
4. Immediate stream reach—reasonably straight, narrow, free of rapids, pools, islands, and eddies; uniform, laminar flow.

5. Obstructions—free of large rocks, bottom debris, algae, weeds, and hanging plants from banks.
6. Bed material—avoid mud, large cobbles, and boulders.

Site Preparation

1. If necessary, move small obstructions and debris (before measurement only).
2. Set tape measure and stakes perpendicular to flow. If impossible, note the correction factor in field notes.
3. Calculate number and size of intervals; recommend 10 percent of flow in any one interval.
4. Find edge of water by looking straight down on the tape measure.
5. Prepare field notes by filling out appropriate blanks; always complete: observer name, date, time, stream name, and stage.

Flow Measurement and Recording

1. For each interval, read and record:
   (a) Distance—from reference point (note edge of water at beginning and end).
   (b) Width—distance halfway from previous to subsequent interval.
   (c) Depth—of water to normal stream bottom.
   (d) Observation depth—distance from water surface to proper meter setting, 0.60 for one velocity measurement or 0.20 and 0.80 for two velocity measurements.
   (e) Meter revolutions—if slow, count complete clicks; if fast, count multiples or use direct reading instrument.
   (f) Time—use a minimum of 40 seconds.
2. Edge effect—watch for partial intervals and calculate appropriately; do not count areas of zero velocity.
3. Calculate discharge—compare to past measurements and site familiarity for reasonableness.

Physical Parameters

TEMPERATURE

Modern pH, conductivity, and dissolved-oxygen meters have temperature probes as part of the instrument. These should be used

All observations, field notes, and field analysis results should be recorded promptly
in the field. Sample labels should be recorded in the field notes. The same field note-
book should be used to allow for comparisons to previous sampling trips.

per manufacturer's instructions. Stream temperatures should be
measured in the stream and not from a subsample unless this is
logistically difficult. Thermometers should be alcohol-based and
allowed to equilibrate in a representative section of the stream.
Temperatures can be recorded to the nearest 0.5°C. Mercury-based
thermometers should be avoided because of often sluggish responses
and because of the potential of thermometer breakage and subsequent release of mercury into the waterbody.

pH

Laboratory procedures for the measurement of pH have been
covered (Chapter 4). Field procedures are the same. The sample used
for the pH measurement should be discarded after use and not
treated as part of the sample to be returned to the laboratory.

A subsample of the stream may be taken for the pH measurement
if the chance of sample warming is small, such as a small difference

between the air and water temperatures. Otherwise the measurement should be taken in the stream. If the water has a low ionic strength, the pH reading may have considerable drift in the moving water. This problem may be alleviated by placing a beaker (nalgene or polyethylene to prevent breakage) underwater upright in the stream. The probe is inserted in the beaker and a measurement is taken. If there is still drift in the reading, the pH reading should be taken after stabilization, or simply recorded after 2 minutes.

CONDUCTIVITY

Laboratory procedures for conductivity measurements have been covered (Chapter 5). Similar techniques are used for the field. Probe placement is the same as pH measurement. Manufacturer's instructions should be consulted.

DISSOLVED-OXYGEN CONTENT

Dissolved-oxygen concentrations vary considerably with depth, temperature, flow rate, time, and other natural factors. Several samplings at different sites and depths may be required for representative values. Laboratory procedures for dissolved oxygen will be covered (Chapter 16). Again, refer to probe placement in the pH section and the specific instrument manufacturer's instructions.

Sample Transport

Sampling procedures were detailed in Chapter 2. Sample bottles should be labeled with permanent waterproof ink on labeling tape. The full sample bottles should be stored in the upright position packed in cold packs, ice, or snow in a cooler with a lid to keep out the light. The samples should be cooled to approximately 4°C. If 4°C is not achievable, the samples should be kept as cool as possible and in the dark until return to the laboratory.

Chapter 7

Gravimetric Analysis—Solids Determination

Purpose: To illustrate the various operations involved in gravimetric analysis, to determine the various categories of solids that are commonly defined in water and wastewater analyses, and to investigate the types of materials that these solids categories define.

Gravimetric analysis is based on determination of constituents or categories of materials by a measurement of their weight. Filtration is used to separate "suspended" or "particulate" (nonfiltrable) fractions from "dissolved" or "soluble" (filtrable) fractions. Filters are generally divided into two categories: pore filters and depth filters. The former remove particles by exclusion from pores of a closely controlled diameter (e.g., Nucleopore membrane filters), while the latter rely on interception of particles that penetrate a deep mat of fibers (e.g., glass and paper filters) (Jenkins *et al.*, 1980).

Evaporation separates water from material dissolved or suspended in it. Analytically, water can be categorized as free and bound; the latter is associated with solids as water of crystallization or as water occluded in the interstices of crystals. Evaporation of water and wastewater samples is normally conducted at two temperatures, 103–105°C and 180°C. The lower temperature is usually used with samples containing high concentrations of organic matter that may suffer significant weight loss due to volatilization and decomposition at the higher evaporation temperature. There is only slight thermal decomposition of inorganic salts at 103°C. Some loss of CO_2 can be expected from the conversion of bicarbonate to carbonate during the dehydration process. Occluded and bound water is not completely removed at 103°C; however, its removal is virtually complete at 180°C (Jenkins et al., 1980).

The analytical objective of combustion in solids determinations is to differentiate between organic and inorganic matter. Organic matter will be destroyed completely by burning at 550°C for 30 minutes.

Procedure

SAMPLING TECHNIQUE

For streams with a stable cross section and uniform lateral distribution of sediments, a single vertical sample will be adequate. Composite samples or averaging is recommended when the cross section is variable. The standard procedure for suspended-sediment sampling is to use the USDI Geological Survey-designed sampler (see p. 41) known as DH-48 (D, depth-integrated; H, hand-held) and designed in 1948 (see Figure 7.1). Insert a clean bottle in the DH-48 and inspect the nozzle for debris. For a depth-integrated sample, hold the sampler just above the water surface pointed directly into the flow. Slowly and evenly, lower the sampler to the streambed and immediately reverse direction to raise the sampler at the previous, even speed. The bottle should be between two-thirds and three-fourths full (do not overfill) or about 350–400 mL. Start over if bottle overfills or is only half full. In low-velocity water, several sampling dips to the streambed may be necessary. Label sample with the following: observer's name, date, time, stream name, and stage (or discharge) (see Figure 7.2). Store in a cold (but not freezing), dark place if not immediately shipped to a lab.

ANALYTICAL TECHNIQUE

Total suspended solids (TSS) can be measured using a paper filter, with evaporation of free water at 103°C. Replicate analyses should be done to maintain quality control.

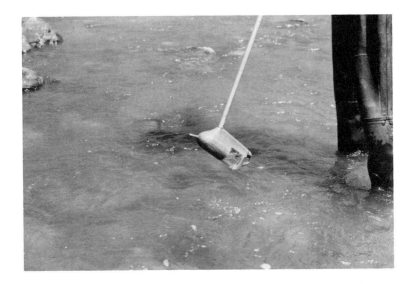

Figure 7.1. The U.S. Geological Survey hand-held depth-integrating sampler (DH-48). Depth-integrated samples should be used when the water depth is over 30 cm, otherwise a grab sample can be used.

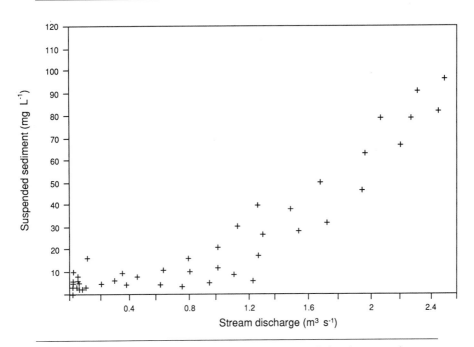

Figure 7.2. A flow-driven relationship between suspended sediment and stream discharge. As stream discharge increases, the suspended sediment concentration increases.

The following apparatuses are used:

1. Glass fiber filters without organic binder or high-quality paper filter.
2. Filtration apparatus.
3. Drying oven (103–105°C).
4. Desiccator.
5. Analytical balance, 200 g capacity, capable of weighing to 0.1 mg.
6. Evaporating dishes and/or weighing pans.
7. Forceps and spatula for handling filters.
8. Sample storage bottles of resistant glass or plastic; suspended material should not adhere to surfaces.

The following procedure can be used to measure TSS dried at 103–105°C:

1. Place filter on filtering apparatus and wash with three successive 20-mL portions of distilled water.
2. Place in aluminum (or other) weighing pan.
3. Dry filter in oven at 103–105°C.
4. Cool filter and weighing pan in desiccator and weigh immediately before using.
5. Pass a relatively large volume of sample (100–300 mL) through the filter to secure a weighable residue.
6. Wash filter and apparatus with three successive portions of 10 mL of distilled water.
7. After filtration, transfer the filter with its contents to the weighing pan and return to an oven maintained at a temperature of 103–105°C and dry for 24 hours.
8. Cool the filter and weighing pan in a desiccator.
9. Weigh the filter and weighing pan.
10. Repeat drying cycle until a constant weight is attained or until weight loss is less than 4 percent of previous weight or 0.5 g, whichever is less.
11. Report the weight over that of the empty filter as suspended solids (nonfiltrable residue) on drying at 105°C in terms of milligrams per liter and report to the nearest whole number. Also report the type of filter used.
12. Calculation:

$$\text{mg L}^{-1} \text{ suspended solids} = \frac{A(1000)}{\text{mL sample}}$$

where A is measured weight of suspended solids in milligrams.

The following procedure can be used to measure total dissolved solids (TDS) dried at 103–105°C:

1. Dry clean crucible in oven at 103–105°C.
2. Cool in desiccator and weigh the crucible.
3. Add 100 mL of filtrate from the suspended solids test to the crucible.
4. Transfer the crucible and its contents into the oven at 103–105°C for 24 hours.
5. Cool the crucible in desiccator.
6. Weigh the crucible.
7. Repeat drying cycle as for suspended solids.
8. Report the weight over that of the empty crucible as total dissolved solids (filtrable residue) on drying at 103–105°C in terms of milligrams per liter and to the nearest whole number. For results exceeding 1000 mg L^{-1}, report only three significant figures. Report the type of filter used.
9. Calculation:

$$\text{mg } L^{-1} \text{ TDS} = \frac{A\ (1000)}{\text{mL of sample}}$$

TYPICAL VALUES

	Total Dissolved Solids (mg L^{-1})	Total Suspended Solids (mg L^{-1})
Cache la Poudre River at canyon mouth near Ft. Collins, Colorado	64.7	16.2
Tualatin River at West Linn, Oregon	91.0	22.7
Rogue River near Agness, Oregon	74.1	22.6
Colorado River at Stateline, Colorado	91.1	240
Colorado River at Imperial, California	74.3	18.4
Skagit River at Marblemount, Washington	31.7	12.4
Yellowstone River near Livingstone, Montana	17.3	47.4
Arkansas River near Coolidge, Kansas	2410	484
Cumberland River near Grand River, Kentucky	106	28.5
Ogeechee River near Eden, Georgia	55	14.3
North Fork Whitewater River near Elba, Minnesota	296	28.9

Review

KEYWORDS

gravimetric analysis	bound water
solids determination	combustion
suspended solids	total suspended solids
soluble solids	total dissolved solids
occluded water	

STUDY QUESTIONS

1. Why are water samples evaporated at two different temperatures?
2. Why are replicates important in the TSS analysis?
3. Given the following TSS data set:

   12, 13, 15, 15, 17, 18, 20, 21, 25, 110

   Calculate the mean, standard deviation, and coefficient of variation. Next calculate the median. Does the mean or median represent the data set?
4. Calculate the percent difference for each replicated analysis. What measures can be taken to decrease any differences?
5. What are some of the possible errors in the TSS analysis? Are they systematic or random errors?

SAMPLE DATA SHEET

Analysis: Gravimetric   Instrument name:

Analyst: Instrument serial number:

Date: Calculations done by:

Start time: Calculations checked by:

  End: Comments:

Procedure:

SAMPLE IDENTIFICATION:	REPLICATE	MEASUREMENT
Standard	1	
	2	
Reported Value:	3	
	Mean:	
	Standard Deviation:	
	In Control?	

SAMPLE IDENTIFICATION:	REPLICATE	MEASUREMENT
	1	
	2	
Reported Value:	3	
	Mean:	
	Standard Deviation:	
	In Control?	

SAMPLE IDENTIFICATION:	REPLICATE	MEASUREMENT
	1	
	2	
Reported Value:	3	
	Mean:	
	Standard Deviation:	
	In Control?	

Suggested Readings

Anderson, H.W., 1970. Relative contributions of sediment from source areas and transport processes. *In* "Proc. Symposium on Forest Land Uses and Stream Environment." Oregon State University, Corvallis, Oregon. 110 pp.

American Public Health Association, 1985. "Standard Methods for the Examination of Water and Wastewaters." Published by the Amer. Public Health Assoc., Amer. Water Works Assoc., and Water Poll. Control Fed., 16th Edition. pp. 133–140.

Beschta, R.L., 1978. Long-term patterns of sediment production following road construction and logging in the Oregon Coast Range. *Water Resources Research* **14(6)**: 1011–1016.

Bilby, R.E., K. Sullivan, and S.H. Ducan, 1989. The generation and fate of road surface sediment in forested watersheds in southwestern Washington. *Forest Science* **35(2)**: 453–468.

Brown, G.W., 1983. "Forestry and Water Quality." (2nd edition). Oregon State University Book Stores, Inc., Corvallis, Oregon. 142 pp.

Brown, G.W., and J.T. Krygier, 1971. Clearcut logging and sediment production in the Oregon Coast Range. *Water Resources Research* **7(5)**: 1189–1199.

Byron, E.R., and C.R. Goldman, 1989. Land-use and water quality in tributary streams of Lake Tahoe, California-Nevada. *J. of Environmental Quality* **18**: 84–88.

Fredriksen, R.L., D.G. Moore, and C.A. Norris, 1975. The impact of timber harvest, fertilization, and herbicide treatments on stream water quality in Western Oregon and Washington. *In* "Forest Soils and Forest Management." Proc. Fourth N. American Forest Soils Conf., B. Bermier and C.H. Winzet (Eds.). Univ. Laval Press, Quebec. pp. 283–313.

Heede, B.H., M.D. Harvey, and J.R. Laird, 1988. Sediment delivery linkages in a chaparral watershed following a wildfire. *Environmental Management* **12(3)**: 349–358.

Jenkins, D., V.L. Snoeyink, J.F. Fergusa, and J.O. Leckie. 1980. "Laboratory Manual for Water Chemistry." Third Edition. Wiley, New York. 183 pp.

Megahan, W.F., and W.J. Kidd, 1972. *Effect of logging roads on sediment production rates in the Idaho Batholith.* USDA Forest Service Inter. For. and Range Exp. Sta. Res. Paper INT-123.

Paustian, S.J., and R.L. Beschta, 1979. The suspended sediment regime of an Oregon Coast Range stream. *Water Res. Bull.* **15(1)**: 144–154.

Wischmeier, W.H., and D.D. Smith, 1978. *Predicting rainfall erosion losses—a guide for conservation planning.* USDA Agric. Handbook No. 537. 58 pp.

Chapter 8

Turbidity

Purpose: To determine the relationship between concentration of different types of particles and turbidity.

When light interacts with a suspension of particles in an aqueous solution, several phenomena occur. Light may interact with the electrons of a particle and be reemitted (scattered) in various directions with the same wavelength as the incident light. Some light may be emitted with a longer visible wavelength than the incident light; some energy may be emitted entirely as long wavelength radiation (heat). Absorption and reemission of light also may occur with dissolved species. The types of interaction that occur between light and particles depend on particle size, particle shape, and the wavelength of the incident light (Jenkins et al., 1980).

Turbidity is defined broadly as the optical property of a suspension that causes light to be scattered rather than transmitted through the suspension. The interactions in natural suspensions are complex and the term "turbidity" is frequently used in a qualitative sense. One definition is "the inverse of that length of solution which will, by scattering, reduce the intensity of a beam of light to 1/2.178 (1/e) of its incident value."

With increasing particle size above a diameter of 5 percent of the wavelength, the intensity of scattered light is a complex function of particle size and angle of light emission. For a constant number of particles and a given wavelength, the scattered light intensity increases to a maximum and then decreases as particle size increases.

Theory of Light Scattering

The optical property expressed as turbidity is the interaction between light and suspended particles in water. A directed beam of light remains relatively undisturbed when transmitted through absolutely pure water, but even the molecules in a pure fluid will scatter light to a certain degree. In samples containing suspended solids, the manner in which water interferes with light transmittance is related to the size, shape, and composition of the particles in the water and to the wavelength (color) of the incident light (Hach et al., 1982).

A minute particle interacts with incident light by absorbing the light energy and then, as if a point light source itself, reradiating the light energy in all directions. This omnidirectional reradiation constitutes the "scattering" of the incident light. The spatial distribution of scattered light depends on the ratio of particle size to the wavelength of incident light (Hach et al., 1982):

1. Particles much smaller than the wavelength of incident light exhibit a fairly symmetrical scattering distribution with approximately equal amounts of light scattered both forward and backward.
2. As particle size increases in relation to wavelength, light scattered from different points of the same particle create interference patterns that are additive in the forward direction. This constructive interference results in forward-scattered light having a greater intensity than light scattered in other directions.
3. In addition, smaller particles scatter shorter (blue) wavelengths more intensely while having little effect on longer (red) wavelengths. Conversely, larger particles do not affect short (blue) wavelengths as much as they scatter long (red) wavelengths.

Particle shape and refractive index also affect scatter distribution and intensity. Spherical particles exhibit a larger forward–backward scatter ratio than do coiled or rod-shaped particles. The refractive index of a particle is a measure of how it redirects light passing through it from another medium such as the suspending fluid. The refractive index of the particle must be different from that of the sample fluid in order for scattering to occur. As the difference between the refractive indices of suspended particle and suspending fluid increases, scattering becomes more intense.

The color of suspended solids and sample fluid are significant in scattered-light detection. A colored substance absorbs light energy in certain bands of the visible spectrum, changing the character of both transmitted light and scattered light and preventing a certain portion of the scattered light from reaching the detection system.

Light scattering intensifies as particle concentration increases. But as scattered light strikes an increasing number of particles, multiple scattering occurs and absorption of light increases. When particulate concentration exceeds a certain point, detectable levels of both scattered and transmitted light drop rapidly, marking the upper limit of measurable turbidity. Decreasing the path length of light through the sample reduces the number of particles between the light source and light detector and extends the upper limit of turbidity measurement.

Two instrumental methods can be used for measuring turbidity. The intensity of the transmitted light beam can be measured and compared with the intensity of the incident light beam. Alternatively, the intensity of the scattered beam can be measured (nephelometric methods). For dilute suspensions, the relationship between the intensity of transmitted light and particle concentration has the form of the Beer–Lambert law if no changes occur in the nature of the suspension as the concentration of particles changes. As the number of particles increases, the possibility of multiple scattering of light also increases thus causing deviations from the Beer–Lambert type of relationship (Hach et al., 1982).

For very dilute suspensions, the decrease in intensity of the light as it passes through the suspension is very small; thus determination of turbidity by transmitted light measurement is not a sensitive procedure. The more sensitive method of determining turbidity is to measure scattered light. This measurement is usually made at right angles to the incident beam.

A visual method for measuring turbidity (the Jackson candle turbidimeter) relies on the observation of the disappearance of the clear image of a candle flame through a column of sample. The measure is dependent on both absorption and scattering of the light entering the sample as well as the characteristics of the human eye. The unit of measurement from this method is the Jackson Turbidity Unit (JTU) and is related only conceptually to the previous rigorous definition of turbidity.

The Jackson candle turbidimeter is designed to measure turbidities between 25 and 1000 units. Solutions having turbidity greater than this must be diluted; solutions with turbidities lower than this are best analyzed by a light-scattering instrument such as the Hach turbidimeter or a spectrophotometer with a nephelometric attachment (device to measure scattered light). The Jackson candle turbidimeter is standardized either with dilutions of a turbid suspension of the same natural water that will be tested or by using a kaolin suspension. Transmission and nephelometric instruments are standardized with a standard suspension of the polymer formazin. The nephelometer turbidity unit is the "Nephelometric Turbidity Unit" or (NTU). It should be noted that there is no reason for any fundamental relationship to exist between turbidity values of a suspension measured by any of the three techniques—visual, transmission, and nephelometric. Nevertheless, empirical correlations can be made over certain ranges of turbidity and for individual suspension.

Procedure

The procedure for the Hach 2100A turbidimeter is presented. This instrument passes a beam of light upward through the sample, where suspended particles scatter an amount of light proportional to the turbidity. A photomultiplier tube located at a 90° angle to the beam then converts the light energy to an electrical signal that is measured on the panel meter in NTUs. Repeat the entire procedure three times for each sample.

Apparatus

1. Hach Model 2100A turbidimeter, including four standard cells (0.61, 10, 100, and 1000 NTU) and four sample cells.
2. Formazin standard suspension (optional).

3. Filtration equipment (extremely turbid samples only).
4. Sample bottles of glass or plastic; suspended material should not adhere to surfaces.

Initiation

1. Before turning instrument on, calibrate the meter to zero with small screw on faceplate.
2. Turn power on. For maximum accuracy, this should be done 12 hours prior to standardizing and running samples.
3. Check focusing by inserting and looking through template in the cell holder. The image of the lamp should just fill the inside circle. If it is off center or too large or small, adjustment can be made according to the procedure in the instrument manual.
4. Calibrate the instrument using Formazin suspensions prepared as described in Standard Methods or the instrument manual or with sealed standard cells provided (0.61, 10, 100, and 1000 NTU). These standards are secondary to the Formazin and should be replaced if flocculation has occurred. Select the desired range, insert the appropriate standard, and adjust STANDARDIZE control until value on meter equals the standard NTU value. When using 100 and 1000 NTU ranges, the cell riser must be used. For more detail, read the following section on operation before calibrating.

Operation

1. Use only clean sample cells with no scratches or etches, fill with 25 mL of sample, and wipe outside of cell clean.
2. Check to be sure air bubbles are not present, place in cell holder, and cover with light shield.
3. Choose sensing range. When using 100- and 1000-NTU ranges only, the cell riser must be placed in cell holder before the sample cell. This decreases the light path length, thus improving linearity of measurements. Do not use in the three lower ranges.
4. Read turbidity in NTUs from scale. When reading 0.2-NTU scale, subtract 0.04 NTU from the reading to account for stray light, which is significant only at this low range.
5. Extremely turbid samples may require dilution with another portion of the same sample that has been filtered. Use of distilled or deionized water may dissolve some of the turbidity. The reading of the diluted sample must then be multiplied

by the dilution factor to obtain turbidity of the original sample. This dilution procedure may need to be repeated if the sample is still too turbid.

6. Remove sample cell, discard sample, and clean sample cell.

Shutdown

1. Remove sample from cell holder.
2. Close sample compartment door.
3. Leave range switch in 100- or 1000-NTU position.
4. Do NOT turn instrument off. It works best when allowed to run continuously.

TYPICAL VALUES

	Turbidity (NTU)
Cache la Poudre River at canyon mouth near Fort Collins, Colorado	8
Tualatin River at West Linn, Oregon	11
Rogue River near Agness, Oregon	7
Colorado River at Stateline, Colorado	290
Colorado River at Imperial, California	12
Skagit River at Marblemount, Washington	3
Yellowstone River near Livingstone, Montana	5
Arkansas River near Coolidge, Kansas	78
Cumberland River near Grand River, Kentucky	20
Ogeechee River near Eden, Georgia	6
North Fork Whitewater River near Elba, Minnesota	12

Review

KEYWORDS

turbidity	JTU
incident light	Nephelometric Turbidity Unit
light reflection	NTU
transmitted light	formazin
Jackson Turbidity Unit	

STUDY QUESTIONS

1. Describe conditions under which turbidity measurements might be correlated with suspended solids measurements. When would there be no correlation?
2. Why is turbidity measured as light refraction and not as light penetration?
3. How would it be possible to have turbidity and yet no suspended solids?
4. Why is an integrated sample best for turbidity measurements?
5. Describe some type of predictive equation using turbidity. Would turbidity be related to stream discharge?

SAMPLE DATA SHEET

Analysis: Turbidity   Instrument name:

Analyst: Instrument serial number:

Date: Calculations done by:

Start time: Calculations checked by:

  End: Comments:

Procedure:

SAMPLE IDENTIFICATION:	REPLICATE	MEASUREMENT
Standard	1	
	2	
Reported Value:	3	
	Mean:	
	Standard Deviation:	
	In Control?	

SAMPLE IDENTIFICATION:	REPLICATE	MEASUREMENT
	1	
	2	
Reported Value:	3	
	Mean:	
	Standard Deviation:	
	In Control?	

SAMPLE IDENTIFICATION:	REPLICATE	MEASUREMENT
	1	
	2	
Reported Value:	3	
	Mean:	
	Standard Deviation:	
	In Control?	

Suggested Readings

American Public Health Association, 1985. "Standard Methods for the Examination of Water and Wastewaters." Published by the Amer. Public Health Assoc., Amer. Water Works Assoc., and Water Poll. Control Fed., 16th Edition. pp. 133–140.

Beschta, R.C., 1980. Turbidity and suspended sediment relationships. In "Watershed Management Symposium." ASCE. Boise, Idoho. pp. 271–282.

Bilby, R.E., K. Sullivan, and S.H. Duncan, 1989. The generation and fate of road surface sediment in forested watersheds in southwestern Washington. *Forest Science* **35(2)**: 453–468.

Brown, G.W., 1983. "Forestry and Water Quality." (2nd Ed.) Oregon State University Book Stores. Inc., Corvallis, Oregon. 142 pp.

Byron, E.R., and C.R. Goldman, 1989. Land-use and water quality in tributary streams of Lake Tahoe, California-Nevada. *J. of Environmental Quality*, **18**: 84–88.

Corbett, E.S., J.A. Lynch, and W.E. Sopper, 1978. Timber harvesting practices and water quality in the eastern United States. *J. Forestry* **76(8)**: 484–488.

Hach, C.C., R.D. Vanous, and J.M. Heer, 1982. *Understanding turbidity measurements*. Hach Technical Information Series, Booklet No. 11. 11 pp. Ames, Iowa.

Jenkins, D., V.L. Snoeyink, J.F. Ferguson, and J.O. Leckie, 1980. "Water Chemistry: Laboratory Manual." (3rd Ed.) 183 pp. Wiley, New York.

Kunkle, S.H., and G.H. Comer. 1971. Estimating suspended sediment concentrations in streams by turbidity measurements. *J. Soil and Water Conserv.* **26(1)**: 18–20.

Chapter 9

Spectrophotometer and Beer's Law

Purpose: To understand the operation of a spectrophotometer and how Beer's Law applies to a particular solution.[1]

Optical methods of analysis can be used to determine satisfactorily the concentration of many dissolved substances. The law relating the amount of light transmitted by a solution to the concentration of a light absorbing constituent is the Beer-Lambert law, better known as "Beer's law":

$$\log (I_0/I) = A = abC$$

where I = intensity of monochromatic light transmitted through the test solution; I_0 = intensity of light transmitted through the reference solution—"the blank"; A = absorbance (dimensionless); b = light path length (usually in centimeters); a = absorptivity, a constant for a given solute–system and a given wavelength; and C = concentration of solute (g L^{-1}).

---

[1] Adapted from Jenkins et al., 1980.

Beer's law states that for a given solution, absorbance is directly proportional to light path length and concentration of absorbing substance. In the instrumental measurement of color intensity, the light transmitted through the solution is measured. The transmittance of a solution (T) is defined as I/I_0, and $\%T$ as $I/I_0 \times 100$.

The colorimeter, or visible wavelength spectrophotometer, is an instrument that makes it possible to quantitatively measure light passing through a clear solution. This instrument is capable of supplying light with a narrow-wavelength bandwidth and is equipped with light-sensitive phototubes to measure light intensity. The light source for an instrument such as the Bausch & Lomb Spectronic 20 is a tungsten lamp. Light from this is dispersed by a diffraction grating (or prism) and the desired wavelength region is selected by passing through a slit. This system produces a wavelength band width of about 20 nanometers (nm). The selected wavelength band is then passed through the sample solution. That light that is not absorbed by the solution is received by the phototube, the signal from which is displayed on the instrument scale. The light source does not emit the same light intensities at all wavelengths, and the phototube is not equally responsive at all wavelengths. Additionally, the solution medium and the cuvette may absorb light of certain wavelengths. The intensity control dial on the spectrophotometer is used to compensate for such effects.

The sensitivity of analysis for a particular solution and the degree of adherence to Beer's law depends on the choice of wavelength. Wavelength selection is based primarily on an evaluation of the solution absorption characteristics as a function of wavelength. Maximum sensitivity, or the largest absorptivity (a), is found at wavelengths where maximum light absorption occurs. Similarly, minimum absorptivities are found where minimum light absorption occurs. Since adherence to Beer's law assumed a constant absorptivity and since the bandwidth produced by the Bausch & Lomb instrument is about 20 nm, it is necessary to choose a nominal wavelength where the absorptivity is approximately equal to that of the immediately adjacent wavelengths. Thus, if possible, a wavelength corresponding to a relatively flat portion of the absorption spectrum is generally chosen. Other portions of the spectrum may be used if a reproducible absorbance–concentration curve can be obtained. Remember that choice of wavelength should consider (a) maximum sensitivity (largest absorptivity and (b) a bandwidth of about 20 nm.

Deviations from Beer's law occur for a number of reasons. For example, because a band of wavelengths is passed through the sample, a deviation can occur, particularly if the molar absorptivity for this band of wavelengths differs significantly. This deviation from Beer's law usually becomes larger at higher concentrations. In some instances chemical reactions of solute, temperature effects, stray or scattered light, fluorescence of solutes, and other factors can also cause deviations.

A standard curve must be prepared in order to convert spectrophotometric data to concentration values. To construct a standard curve, prepare a series of dilutions of pure constituent (from reagent) and analyze each solution using the same procedure as will be used for analyzing the samples. A curve of absorbance (or transmission) at a fixed wavelength, versus concentration, can then be drawn. The absorbance of an unknown sample is then compared to the standard curve to find its concentration value. If an estimate of the concentration in the unknown sample is available, it is advisable to prepare the standard curve in the appropriate concentration range. Alternatively, the sample can be diluted to a concentration value within the range of the prepared standard curve.

Absorbance is read by comparing the sample to a "blank" solution. The blank does not contain any of the constituent of interest. A blank always has 100 percent transmittance (= 0 percent absorbance). The blank must receive all the same treatments as the samples to be comparable.

Samples are entered in the spectrophotometer in special "cuvettes" made for the instrument. In order to eliminate interference due to cuvette differences, use only two cuvettes throughout the procedure. One will contain the blank, and the other will be used for samples. After all samples have been analyzed, fill both cuvettes with distilled water. The absorbance from both cuvettes should be 0 percent. However, if the cuvette that held the blank has absorbance greater than the cuvette used for samples, the absorbance difference should be added to sample absorbance values. If the blank cuvette has absorbance less than the sample cuvette, subtract the difference from all sample absorbance values.

Review

KEYWORDS

spectrophotometer
Beer's law
light path length
transmittance
colorimeter
cuvette
wavelength band
absorption spectra
blank

Suggested Reading

Jenkins, D.L., V.L. Snoeyink, J.F. Ferguson, and J.O. Leckie, 1980. "Water Chemistry: Laboratory Manual." Third Edition. 183 pp. Wiley, New York.

Chapter 10

Phosphorus

Purpose: To examine a colorimetric method of analysis and apply it for determination of phosphate in water.

Phosphorus is found in soils, plants, and microorganisms in a number of organic and inorganic forms. It is second only to nitrogen as a nutrient element required by plants and microorganisms. Phosphorus is considered an essential nutrient element because it is required or is favorable for (1) seed formation, (2) root development, (3) maturation of crops, especially cereals, (4) crop quality, (5) strength of straw in cereal crops, and (6) accumulation and release of energy during cellular metabolism.

Under natural conditions the phosphorus taken up by growing plants is returned to soils in animal and vegetative residues and remains. Under cultivation, however, part of the element taken up by the crop is removed in harvests and only part is returned to the soils. To offset removals of phosphorus and to increase levels where they are naturally low, fertilizers are added to soils.

The form in which phosphorus is likely to be present in natural water is somewhat uncertain, but the most probable species would appear to be phosphate anions, complexes with metal ions, and colloidal particulate material. Phosphorus is most often reported in terms of an equivalent amount of orthophosphate (PO_4^{3-}). Like

111

nitrogen, a wide range of oxidation states is possible for phosphorus, but no strong resemblance in aqueous chemical behavior between the two elements is apparent. Concentrations of orthophosphate normally present in natural water are far less than those of nitrate (Hem, 1985).

Sources of Phosphate

The most common rock mineral in which phosphorus is a major component is apatite, a general name for general species that are principally calcium orthophosphate and that also contain fluoride, hydroxide, or chloride ions (Hem, 1985). These minerals are widespread in both igneous rock and in marine sediments. When apatite is attacked by water, the phosphorus species released probably recombine rather rapidly to form other minerals or are absorbed by hydrolyzate sediments, especially clay minerals, in soil. Phosphate is made available for solution in water from several kinds of cultural applications of phosphate in the activities of humans, and these pollution sources probably are the most important causes of high concentrations of phosphate in surface water.

Phosphorus is a component of sewage, as the element is essential in metabolism, and it is always present in animal metabolic waste. The use of sodium phosphate as a binder in household detergents probably has greatly increased the output of phosphate by sewage-disposal plants. Recognition of household detergents being a phosphorus source led to the development of low-phosphate detergents. Only about 30 percent of the phosphate in wastewater comes from human excrement. A detergent molecule works by the hydrocarbon end dissolving in fat (soiled material), while the ionic end dissolves in water. Fats are thus brought into solution. The hydrocarbon portion may be biodegradable, but phosphates are added to tie up calcium and to aid cleaning action and will remain in solution.

Other phosphate sources include phosphate fertilizers. Phosphate fertilizer application, especially when coupled with irrigation, is a potential source of phosphate in drainage waters. Orthophosphate concentrations increase as sediment concentrations increase. The suspended sediment material as colloidal particulates may carry sorbed phosphate. Irrigated agricultural fields may be large non-point-sources of orthophosphate from fertilizers and soil erosion.

Chemistry of Phosphate in Water

The orthophosphate ion (PO_4^{3-}) is the final dissociation product of phosphoric acid, H_3PO_4. The dissociation of the acid occurs in steps, and four solute species are possible: $H_3PO_4(aq)$, $H_2PO_4^-$, HPO_4^{2-}, and PO_4^{3-}. For solutions of low ionic strength, a disssociation diagram can be used to give a reasonable estimate of the proportion of total phosphate as the various phosphate species when pH is known (Hem, 1985).

Phosphate ions form complexes with many of the other solutes present in natural water. A calculation of solubility controls over phosphate activity is difficult when so many solute species containing phosphorus would be possible and the form of solids that might be produced is uncertain. Phosphate contents of natural water are not generally restricted by precipitation of a sparingly soluble inorganic compound. More likely, phosphorus uptake by aquatic vegetation and perhaps the adsorption of phosphate ions by metal oxides, especially ferric hydroxide, are the major factors that prevent concentrations greater than a few tenths or hundredths of a milligram per liter from being present in solution in most waters.

In the process of titrating for alkalinity, all the HPO_4^{2-} will be converted to $H_2PO_4^-$, and that fraction would appear in the alkalinity value as an equivalent quantity of bicarbonate. Refer to Chapter 12 for more details.

Growth of aquatic vegetation, such as algae, may be influenced by the availability of nitrogen and/or phosphorus. Dense, rapidly multiplying algal growths or blooms sometimes occur in water bodies that periodically receive increased concentrations of nitrogen or phosphorus. These dense growths are generally undesirable to water users and may interfere with other forms of aquatic life, especially if the waterbody becomes overloaded with oxidizable debris as a result of the sudden dieback of an algal bloom (Hem, 1985). The large amount of oxygen required in the organic matter decomposition may result in an oxygen deficit.

The enrichment of a waterbody with nutrients is termed "eutrophication" and is accompanied by a high rate of production of plant material in the water. Troublesome production rates of vegetation presumably can only occur when optimum supplies of all nutrients are present and available.

Phosphate Analyses

Phosphate exists in several forms in natural waters. These are broadly categorized as orthophosphate (i.e., PO_4^{3-}, HPO_4^{2-}, condensed phosphates (i.e., pyro-, poly-, and metaphosphates), and organic phosphate. These forms can be present in true solution or as particulates.

Phosphate analyses require two general steps: (1) conversion of the phosphorus form of interest to soluble orthophosphate and (2) colorimetric determination of soluble orthophosphate. In addition, it may be of interest to separate filtrable (dissolved) phosphate species from nonfiltrable (particulate) phosphate before conversion to soluble orthophosphate. If this is the case, filtration of the sample through a 0.45μm membrane filter precedes the sample conversion to orthophosphate. The nonfiltrable fraction may then be determined by difference (total filtrable and nonfiltrable phosphate in the form of interest minus filtrable phosphate in the same form).

To determine orthophosphate, colorimetric determination is performed on a sample without any preliminary conversion. Colorimetric determination is accomplished in four different ways, three of which are (1) the vanadomolybdate method, (2) the stannous chloride method, (3) the ascorbic acid method (APHA, 1985), and (4) the amino acid method available as a commercial kit (Hach Chemical C(1981).

To conve condensed phosphates (pyro-, poly-, and metaphosphates) to orthophosphate, acid hydrolysis precedes colorimetric determination. For this reason, condensed phosphate plus orthophosphate measurement is known as "acid-hydrolyzable" phosphate. The condensed phosphate portion is determined by subtracting total orthophosphate from total acid-hydrolyzable phosphate.

Organic phosphate is converted to orthophosphate by digestion or oxidation of the organic water by digestion, listed in order of decreasing rigor: (1) perchloric acid digestion, (2) sulfuric acid–nitric acid digestion, and (3) persulfate digestion. The rigor of the digestion required is dependent on the type of sample. After digestion, colorimetric analysis determines total phosphate content (organic, plus condensed, plus orthophosphate). The organic portion is determined by subtracting total acid-hydrolyzable phosphate from total phosphate.

Important note: When doing phosphate analyses, all glassware must be acid washed with dilute HCl to remove adsorbed phosphate left by commercial detergents. After the acid washing, rinse well with deionized (DI) water; distilled water (DW) may still have the potential to contaminate the glassware.

Procedure

ORTHOPHOSPHATES

Three different methods of orthophosphate determination may be used: ascorbic acid, amino acid, and vanadomolybdate methods.

Ascorbic Acid Method (Murphy–Riley) (APHA *et al.*, 1985)

Natural color of water generally doesn't interfere with this test because of the high wavelength used.

Apparatus

1. Spectrophotometer, set at 885 nm.
2. Graduated cylinder.
3. Several 150 mL beakers.
4. Automatic pipettor.

Procedure

1. Measure a 50 mL sample into a 150 mL beaker.
2. Add 5 mL of Armstrong's reagent and 1.0 mL ascorbic acid, swirl to mix.
3. Allow at least 20 minutes for color development, but not longer than 2 hours before reading.
4. Adjust the spectrophotometer to 0 absorbance with a portion of untreated sample.
5. Now read the absorbance of your sample and compare the result to the standard curve.

Special reagents (APHA *et al.*, 1985) are as follows:

1. Armstrong reagent: Add 122 mL of concentrated H_2SO_4 to 800 mL of DI water. While the solution is still hot, add 10.5 g of ammonium molybdate and 0.3 g of antimony potassium titrate.

Heat to dissolve; cool, and dilute to exactly 1 L with DI water, stable indefinitely.

2. Ascorbic acid solution: Dissolve 3 g of ascorbic acid in 100 mL of DW. *Note:* Stable only one week (store in refrigerator).

Amino Acid Method (Hach)

Apparatus

1. Spectrophotometer, set at 700 nm.
2. Graduated cylinder.
3. Several 50 mL Erlenmeyer flasks.
4. Several rubber stoppers.
5. Automatic pipettor.

Procedure

1. Measure a 25 mL sample into a 50 mL graduated cylinder.
2. Add 1.0 mL of molybdate reagent (Hach).
3. Add contents of one amino acid powder pillow or 1 mL of amino acid reagent solution.
4. A blue color indicates presence of phosphate. Wait 10 minutes for full color development, but no longer than 15 minutes before reading.
5. Adjust 0% transmittance (T) and then adjust 100% T with a portion of 25 mL DI water plus all reagents.
6. To find mg L^{-1} phosphate, use a standard curve.

For amino acid, possible interferences and what to do are as follows. Very high phosphate concentrations will cause a green, yellow, or brown color. If you get one of these colors, try a sample dilution.

Vanadomolybdate Method

Apparatus

1. Spectrophotometer, set at 440 nm.
2. Graduated cylinder.
3. Several 50 mL Erlenmeyer flasks.
4. Rubber stoppers.
5. Pipettes.

Procedure

1. Set 100% T using a DW blank.
2. Pipette 5 mL sample into a 50 mL Erlenmeyer flask. Add 10 mL of vanadomolybdate reagent (dissolve 25 g ammonium molybdate in 300 mL DW) and dilute to 50 mL with distilled H_2O using a graduated cylinder. Mix thoroughly and wait 10 minutes for color development.
3. Measure absorbance at both 440 and 490 nm. Be sure to check the 100% T settings each time the wavelength is changed.
4. Compare the results to the standard curves.

TOTAL INORGANIC

In this test, all metaphosphate is converted to orthophosphate, which is then measured directly. See step 6 below for replication instructions.

To determine the concentration of metaphosphate, a test is run for total inorganic phosphate (ortho plus meta), and the results of the orthophosphate test are subtracted from the results of this test. The difference is the concentration of metaphosphate.

Apparatus

All apparatus used in both orthophosphate tests plus a hotplate.

Procedure

1. Measure two separate 25 mL samples into separate 50 mL Erlenmeyer flasks.
2. Add 2.0 mL H_2SO_4 solution to each and swirl to mix.
3. Place both samples on a hotplate and boil gently for 30 min. Maintain the sample volume at 20 mL with DI water.
4. Allow samples to cool, then add 2.0 mL of 1 N NaOH solution to each, swirl to mix, and return pH to phenothalein endpoint.
5. Pour samples into separated graduated cylinders and return sample volume to 25 mL with DI water.
6. Run one sample for orthophosphate using the "Ascorbic Acid" procedure (use only half volumes of each reagent since sample size is half volume). Run another sample for orthophosphate

using the "Amino Acid" procedure. Run the third sample using the "Vanadomolybdate" method.

7. The mg L^{-1} metaphosphate equals the mg L^{-1} total phosphate minus the mg L^{-1} orthophosphate.

TYPICAL VALUES

	Orthophosphate (mg L^{-1} as PO_4)
Cache la Poudre River at canyon mouth near Ft. Collins, Colorado	0.03
Tualatin River at West Linn, Oregon	0.66
Rogue River near Agness, Oregon	0.05
Colorado River at Stateline, Colorado	0.07
Colorado River at Imperial, California	0.05
Skagit River at Marblemount, Washington	0.01
Yellowstone River near Livingstone, Montana	0.05
Arkansas River near Coolidge, Kansas	0.12
Cumberland River near Grand River, Kentucky	0.12
Ogeechee River near Eden, Georgia	0.04
North Fork Whitewater River near Elba, Minnesota	0.13

Review

KEYWORDS

colorimetric analysis
phosphorus
organic
inorganic
phosphate
orthophosphate
sewage
detergent
eutrophication
Armstrong's reagent
metaphosphate
total phosphorus

STUDY QUESTIONS

1. Explain how acid rinsing of glassware improves the accuracy of phosphate analyses in water samples.
2. Describe the procedure for preparing phosphate-free glassware.
3. Calculate the percent species of orthophosphate given the dissociation of H_3PO_4 at pH 7.

$$H_3PO_4 \rightleftharpoons H_2PO_4^- + H^+$$

$$H_2PO_4 \rightleftharpoons HPO_4^{2-} + H^+$$

$$HPO_4 \rightleftharpoons PO_4^{3-} + H^+$$

4. Report your estimate of the condensed phosphate and orthophosphate content of the samples provided. Show your calculations.

SAMPLE DATA SHEET

Analysis: Phosphorus
Analyst:
Date:
Start time:
  End:
Procedure:

Instrument name:
Instrument serial number:
Calculations done by:
Calculations checked by:
Comments:

SAMPLE IDENTIFICATION:	REPLICATE	MEASUREMENT
Standard	1	
	2	
Reported Value:	3	
	Mean:	
	Standard Deviation:	
	In Control?	

SAMPLE IDENTIFICATION:	REPLICATE	MEASUREMENT
	1	
	2	
Reported Value:	3	
	Mean:	
	Standard Deviation:	
	In Control?	

SAMPLE IDENTIFICATION:	REPLICATE	MEASUREMENT
	1	
	2	
Reported Value:	3	
	Mean:	
	Standard Deviation:	
	In Control?	

Suggested Readings

American Public Health Association, 1985. "Standard Methods for the Examination of Water and Wastewaters." 16th Ed. Published by the Amer. Public Health Assoc., Amer. Water Works Assoc., and Water Poll. Control Fed. Washington, D.C.

Borman, F.H., G.E. Likens, and J.S. Eaton, 1969. Biotic regulation of particulate and solution losses from a forest ecosystem. *Bioscience* **19(7)**: 600–610.

Brown, G.W., A.R. Gabler, and R.B. Marston, 1973. Nutrient losses after clearcut logging and slash burning in the Oregon Coast Range. *Water Resources Research* **9**: 1450–1453.

Byron, E.R., and C.R. Goldman, 1989. Land-use and water quality in tributary streams of Lake Tahoe, California-Nevada. *J. of Environmental Quality*, **18**: 84–88.

Corbett, E.S., J.A. Lynch, and W.E. Sopper, 1978. Timber harvesting practices and water quality in eastern United States. *J. Forestry* **76(8)**: 484–488.

Gessel, S.P., and D.W. Cole, 1965. Influences of removal of forest cover on movement of water and associated elements through soil *J. Amer. Water Works Assoc.*, **57(10)**: 1301–1310.

Hem, J.D., 1985. "*Study and interpretation of the chemical characteristics of natural water.* U.S. Geological Survey Water Supply Paper 2254. Washington, D.C. 263 pp.

Likens, G.E., F.H. Bormann, N.M. Johnson, D.W. Fischer, and R.S. Pierce, 1970. Effects of forest cutting and herbicide application on nutrient budgets in the Hubbard Brook watershed-ecosystem. *Ecol. Mono.* **40(1)**: 24–46.

Novotny, V., and G. Chesters, 1981. "Handbook of Nonpoint Pollution: Sources and Management. "Van Nostrand Reinhold Co., New York. 555 pp.

Chapter 11

Nitrogen

Purpose: To determine the concentration of ammonia (NH_3) as ammonium (NH_4^+), nitrite (NO_2^-), and nitrate (NO_3^-) nitrogen.

A considerable part of the total nitrogen of the earth is present as nitrogen gas in the atmosphere. The oxidation and reduction of aqueous nitrogen species are closely tied to biological activity, and both the paths followed and the end products of such reactions depend very strongly on kinds and numbers of biota present. Small amounts of nitrogen are present in rocks, but nitrogen is concentrated to a greater extent in soil and biological material as organic nitrogen. The influence of biota on the nitrogen reactions and the departure from equilibrium conditions severely limit the value of pH–Eh diagrams in predicting species to be expected in natural water. In general, the oxidation of organic nitrogen in air can be expected to produce nitrite and finally nitrate. Some anaerobic organisms reduce nitrate nitrogen and can produce nitrogen gas instead of, or in addition to, ammonium. In groundwater, nitrate usually appears to be the only form of nitrogen of significance, although nitrite-bearing water may occur in reducing environments (Hem, 1985).

Nitrogen in the form of dissolved nitrate is a major nutrient for vegetation, and the element is essential to all life. Certain species of bacteria in soil, especially those living on roots of legumes, and the blue–green algae, and other microbiota occurring in water, can extract nitrogen from air and convert it into nitrate. Some nitrate occurs in rainwater, although the belief that a significant fraction of this is produced by lightning discharges now seems incorrect, especially given increased NO_x concentrations and acid precipitation formation. Nitrate in the soil that is utilized by plants is partly returned to the soil when the plants die, although some nitrate is lost from the cycle in drainage and runoff and appears in river water.

Evidence of the importance of soil leaching in producing the nitrate concentrations observed in river water can be gathered from records of river-water quality for streams in productive agricultural regions. Farm animals produce considerable amounts of nitrogenous organic waste that tends to concentrate in places where large numbers of animals are confined. The occurrence of high nitrate concentrations in shallow groundwater has been attributed to leachings from livestock wastes from feedlots. Groundwater studies in the South Platte Valley of Colorado showed substantial contributions of nitrogen and phosphorus beneath irrigated fields, and feed lots (Hem, 1985).

Sources of reduced nitrogen forms in natural water presumably are similar to the nitrate sources, and the state of oxidation of nitrogen probably is controlled by biochemical processes. Although the reduced forms normally would be transformed to nitrate in most natural-water environments, there is considerable evidence that a significant amount of reduced nitrogen may be present in many waters.

The pH at which the transformation of aqueous ammonia to ammonium ion is half completed is about 9.24 (Sillen and Martell, 1964). This is above the usual pH of natural water and suggests that in most environments any ammonia nitrogen in solution would have the form NH_4^+ (ammonium). Most of the nitrogen dissolved in rainwater appears to occur in the form of ammonium ions. The circulation in the atmosphere of this form is facilitated by its volatility.

In natural waters, nitrogen exists as nitrate, nitrite, ammonium, and organic nitrogen (in order of decreasing oxidation state). All of the above forms are biochemically interconvertible, forming the nitrogen cycle (with nitrogen gas). Nitrate is found in small amounts in fresh domestic wastewater, but may attain high levels in groundwater. In nature, nitrates are formed by bacteria that change organic nitrogen to nitrites (NO_2^-) and quickly to nitrates (NO_3^-) under aerobic conditions. Nitrate (NO_3^-) nitrogen is the most completely oxidized state of nitrogen found in water.

Nitrate is a required nutrient for algae, and too much can lead to eutrophication of lakes and streams since it is often a limiting nutrient in algal growth. High levels in drinking water can cause infant methemoglobinemia (blue baby syndrome). This has led to the maximum allowable limit of 45 mg L^{-1} nitrate or 10 mg L^{-1} nitrate nitrogen for public drinking-water supplies. Nitrite (NO_2^-) nitrogen occurs as an intermediate stage in biological decomposition of compounds containing organic nitrogen. Bacteria converts ammonium nitrogen to nitrites under aerobic conditions. Under anaerobic conditions, certain bacteria can reduce nitrates to nitrites.

In surface waters, nitrites are quickly oxidized to nitrates, and so their presence indicates partially decomposed organic wastes in the water. Drinking-water supplies rarely contain more than 0.1 mg L^{-1} nitrite.

Total oxidized nitrogen is the sum of nitrite and nitrate nitrogen. The cadmium reduction method tests for both nitrite and nitrate nitrogen. Dilution is necessary since this modification operates in the 0–0.1-mg L^- range of nitrate–nitrogen.

Samples should be stored at 4°C or lower if the sample is to be analyzed within 24 hours. For longer storage periods, further preservation steps should be taken.

The Kjeldahl method determines nitrogen in the trinegative oxidation state. It fails to account for nitrogen in the form of azide, azine, azo, hydrazone, nitrate, nitrite, nitrile, nitro, nitroso, oxime, and semicarbazone (APHA, 1985). If ammonia nitrogen is not removed in the initial phase, the term "Kjeldahl nitrogen" is applied to the result. Organic nitrogen is the sum of organic nitrogen, nitrate, nitrite, and ammonium.

Two of the most common ways that ammonia (or ammonium) can enter a surface-water system are domestic pollution from wastes or runoff from farm fields. Ammonia may be used directly on plants and is a common ingredient in commercial fertilizers.

Procedure

AMMONIA

The two most common methods for ammonia analysis are the Phenate method (for 0.1–0.5 mg L^{-1} NH_4–N) and the Nessler method (for 0.2–5 mg L^{-1} NH_4–N).

Phenate Method

An intensely blue compound, indophenol, is formed by the reduction of ammonia, hypochlorite, and phenol catalyzed by a manganous salt. The deepness of the blue color formed depends on the amount of ammonia present. By comparing the reading of a sample to those obtained with a standard curve, you can determine the concentration of ammonia.

Apparatus

1. Spectrophotometer, set at 630 nm.
2. Graduated cylinder.
3. Several 100-mL breakers.
4. Magnetic stirrer and spinbars.

Procedure

1. Measure a 20-mL sample, pour into a 100-mL beaker, and set to MIXING on magnetic stirrer.
2. Add two drops of $MnSO_4$ solution (dissolve 50 mg $MnSO_4$ in 100 mL DW).
3. Add 1 mL hypochlorous acid solution (NaOCl, DW, and HCl; see Standard Methods).
4. Immediately add 1.2 mL phenate reagent, drop by drop.
5. Wait 10–15 minutes for color development. Adjust spectrophotometer using treated deionized sample. Measure transmittance of sample on spectrophotometer.
6. Use a standard curve to convert from absorbance to milligrams per liter of ammonia.

Nessler Method

This is also a colorimetric analysis, but it uses a yellow–brown color development. It is a much easier method.

Apparatus

1. Spectrophotometer, set at 425 nm.
2. Several 50-mL-Erlenmeyer flasks.
3. Some rubber stoppers.
4. Graduated cylinder.
5. Automatic pipettor.

Procedure

1. Measure a 25-mL sample into a 50-mL Erlenmeyer flask; also measure 25 mL deionized water into a separate flask.
2. Add one drop of Rochelle salt reagent (see Standard Methods) if needed. (This will eliminate interference due to hardness.)
3. Add 1 mL of Nessler's reagent (HgI_2, KI, and NaOH in DW; see Standard Methods) to both flasks, stopper, and invert several times to mix.
4. Allow 10 minutes for color development, but don't wait longer than 25 minutes before taking the reading.
5. Adjust the zero on the spectrophotometer, then place a portion of the treated deionized water in the spectrophotometer and adjust to 100% transmittance (T).
6. Place a portion of the treated sample in the spectrophotometer and read the percent transmittance.
7. To find the concentration (mg L^{-1}) ammonia nitrogen, compare your results with the standard curve.

For phenate, possible interferences and what to do are as follows: More than 500 mg L^{-1} alkalinity, 100 mg L^{-1} acidity, color, and turbidity all interfere, but can all be removed by distillation. For Nessler's reagent, iron and sulfide are most common; also, many types of organic compounds (i.e, hydrazine, acetone, aldehydes, alcohols), can all be removed by distillation.

COLORIMETRIC NITRITE AND NITRATE (HACH)

Hach water analysis methods for determining nitrate concentrations can be used. The Hach field kit for nitrate may also be used. Directions for field determination of nitrate are included with the kit.

Nitrate (Hach)

Apparatus

1. Spectrophotometer, set at 500 nm.
2. Several 50-mL Erlenmeyer flasks.
3. Several rubber stoppers.
4. Graduated cylinder.

Procedure

1. Measure 25-mL sample into a 50-mL Erlenmeyer flask, then measure 25 mL DI water in a separate flask.
2. Add contents of an NitraVer 5 Nitrate Reagent Powder Pillow, stopper, and shake vigorously for 1 minute.
3. Wait at least 5 minutes for color development, but no longer than 15 minutes before reading (amber color indicates nitrate).
4. Adjust the spectrophotometer by zeroing without a sample (0% T), then placing a portion of treated DI water in the spectrophotometer and adjusting to 100% T. The ammonia free sample should be treated with the same reagents as the water sample before adjusting the spectrophotometer.
5. To find the milligrams per liter of total oxidized nitrogen, compare results to a prepared standard curve.

Nitrite (Hach)

Apparatus

1. Spectrophotometer, set at 560 nm.
2. Several 50-mL Erlenmeyer flasks.
3. Several rubber stoppers.

Procedure

1. Measure 25-mL sample into a 50-mL Erlenmeyer flask, then measure 25 mL DI water in a separate flask.
2. Add contents of one NitraVer 2 Nitrite Reagent Powder Pillow and swirl to mix. A greenish-brown color will develop if nitrite nitrogen is present. Allow 10 minutes for the color to fully develop.

3. Read on spectrophotometer at 560 nm.
4. Find the milligrams per liter from the standard curve.

CADMIUM REDUCTION COLUMN FOR NITRATE

The cadmium reduction method is used in two different ways for nitrate analysis. In both procedures the nitrate (NO_3^-) is reduced to nitrite (NO_2^-) and it is this form of nitrogen that is measured. Because of this, it is necessary to subtract any concentration of nitrite found in the last procedure from the final answer.

The first procedure uses a reduction column of amalgamated cadmium filings. The second uses premixed, premeasured Hach reagents, greatly simplifying the analysis (see Hach for procedure).

Apparatus

1. Spectrophotometer, set at 543 nm.
2. Reduction column containing amalgamated cadmium filings.
3. Several 150 mL beakers.
4. Graduated cylinder.
5. pH meter.

Procedure

1. Measure a 100-mL sample into a 150-mL beaker.
2. Check the pH of the sample. If it is greater than 9, adjust to between 8 and 9 with dilute HCl.
3. Add 2.0 mL concentrated NH_4Cl solution, mix, and pour onto column.
4. Collect about 30 mL effluent from the column and discard (the first 30 mL "cleans" out the column).
5. Collect the remainder of the effluent (at least 50 mL) in a beaker.
6. To exactly 50 mL of effluent add 1.0 mL sulfanilamide solution, mix, then wait 3–5 minutes.
7. Add 1.0 mL l-naphthylethylenediamine solution, mix, then wait 10 minutes.
8. Measure the absorbance of the sample against a prepared blank.
9. Use a standard curve to find the milligrams per liter of NO_3–N from absorbance values.

TYPICAL VALUES

Ammonium-nitrogen (NH_4–N) and nitrate–nitrogen (NO_3–N):

	NH_4–N (mg L^{-1})	NO_3–N (mg L^{-1})
Cache la Poudre River at canyon mouth near Fort Collins, Colorado	0.03	0.07
Tualatin River at West Linn, Oregon	0.41	1.47
Rogue River near Agness, Oregon	0.04	0.14
Colorado River at Stateline, Colorado	0.14	0.64
Colorado River at Imperial, California	0.03	0.13
Skagit River at Marblemount, Washington	0.03	0.08
Arkansas River near Coolidge, Kansas	0.09	1.56
Cumberland River near Grand River, Kentucky	0.10	0.42
Ogeechee River near Eden, Georgia	0.02	0.07
North Fork Whitewater River near Elba, Minnesota	0.60	2.28

Review

KEYWORDS

nitrogen NO_x
nitrate   limiting nutrient
nitrite   Nessler
fertilizer   ammonia
NH_4^+ cadmium reduction
NO_3^-

STUDY QUESTIONS

1. Is soil nitrogen in organic or inorganic form? Explain.
2. Name two possible anthropogenic sources of nitrogen in groundwater.
3. Name two naturally occurring processes that produce nitrate.
4. Why are nitrite concentration in unpolluted surface waters invariably low?
5. Define the concept of nutrient conservation.
6. What advantages are there for fertilizing with ammonium N rather than nitrate N?

SAMPLE DATA SHEET

Analysis: Nitrogen

Analyst:	Instrument name:
Date:	Instrument serial number:
Start time:	Calculations done by:
End:	Calculations checked by:
Procedure:	Comments:

SAMPLE IDENTIFICATION:	REPLICATE	MEASUREMENT
Standard	1	
	2	
Reported Value:	3	
	Mean:	
	Standard Deviation:	
	In Control?	

SAMPLE IDENTIFICATION:	REPLICATE	MEASUREMENT
	1	
	2	
Reported Value:	3	
	Mean:	
	Standard Deviation:	
	In Control?	

SAMPLE IDENTIFICATION:	REPLICATE	MEASUREMENT
	1	
	2	
Reported Value:	3	
	Mean:	
	Standard Deviation:	
	In Control?	

Suggested Readings

American Public Health Association, 1985. "Standard Methods for the Examination of Water and Wastewaters." Published by the Amer. Public Health Assoc., Amer. Water Works Assoc., and Water Poll. Control Fed., 16th Ed. pp. 373–412.

Bormann, F.H., G.E. Likens, and J.S. Eaton, 1976. Biotic regulation of particulate and solution losses from a forest ecosystem. *Bioscience* **19(7)**: 600–610.

Brosten, D. 1988. Nitrogen stabilizers show promise. *Agrichemical Age* **32(3)**: 8–16.

Gessel, S.P., and D.W. Cole. 1965. Influences of removal of forest cover on movement of water and associated elements through soil. *J. Amer. Water Works Assoc.* **57(10)**: 1301–1310.

Hach Chemical Co. 1975. "Hach Water and Wastewater Analysis Procedures Manual," 3rd Ed. (2-75)–(2-82).

Hem, J.D. 1985. *Study and Interpretation of the Chemical Characteristics of Natural Water*. U.S. Geological Survey Water Supply Paper 2254. Washington, D.C. 263 pp.

Likens, G.E., F.H. Bormann, N.M. Johnson, D.W. Fischer, and R.S. Pierce. 1970. Effects of forest cutting and herbicide application or nutrient budgets in the Hubbard Brook watershed-ecosystem. *Ecol. Mono.* **40(1)**: 24–46.

Novotny, V., and G. Chesters. 1981. "Handbook of Nonpoint Pollution: Sources and Management." Van Nostrand Reinhold Co., New York. 555 pp.

Sillen, L.G., and A.E. Martell, 1964. Stability constants of metal–ion complexes. Chemical Society [London] Special Publication 17. 754 pp.

Chapter 12

Alkalinity

Purpose: To learn methods and applications for the analysis of alkalinity, and to understand the effects of carbonate equilibria on solutions.

Theory

The alkalinity of a water is the capacity of that water to neutralize acid. Most of the alkalinity in natural waters is contributed by bicarbonate (HCO_3^-), carbonate (CO_3^{2-}), and hydroxide (OH^-) species. Carbon oxides enter into a number of reactions in water, which may be summarized as

$$H_2O(1) + CO_2(g) \leftrightharpoons H_2CO_3 \leftrightharpoons H+ + HCO_3^- \leftrightharpoons H+ + CO_3^{2-}$$

The pH of the water determines the relative concentrations of these carbonate species. Carbonic acid is diprotic and the dissociation of carbonic acid (H_2CO_3) to bicarbonate (HCO_3^-) and to carbonate (CO_3^{2-}) is pH-dependent. Most natural waters have a pH of 5.5–7.5 and HCO_3^- is the dominant carbon species; thus alkalinity is often expressed as the equivalents of HCO_3^-. Most waters have alkalinity in milliequivalents per liter (meg L^{-1}), while dilute waters

Some water-quality analyses may be done quickly in the field with commercially available field test kits. Analyses usually include pH, alkalinity, and dissolved oxygen.

may have alkalinity expressed as microequivalents per liter (μeq L^{-1}). A still common expression is mg L^{-1} of calcium carbonate (CaCO$_3$).

Alkalinity is important in a number of ways. Excessive alkalinity may render water unsuitable for irrigation depending on the nature of the cations (Ca^{2+}, Mg^{2+}, Na$^+$, and K$^+$), or improving the matrix resistance to low pH caused by acid rain or wastewater.

Alkalinity is determined by titration with a standard solution of a strong mineral acid to the successive bicarbonate and carbonic acid equivalence points, indicated electrometrically or by means of a color indicator. The calculated alkalinity depends on the indicator endpoint used, which may not be exactly the same as the true equivalence point. Therefore, it is important to describe the method used when reporting data. The stoichiometric relationships between hydroxide, carbonate, and bicarbonate are valid only in the absence of significant concentration of weak-acid radicals other than hydroxyl, carbonate, or bicarbonate.

There are several different measures of alkalinity. Caustic alkalinity is equivalent to the amount of strong acid required to lower the pH of a sample to (10.5). The phenolphthalein (carbonate) alkalinity is the amount of strong acid moles per liter (mol L^{-1}) required to lower the pH of a sample to (8.3). Fortuitously, the phenolphthalein indicator has an endpoint color change at approximately 8.3, allowing us to measure the alkalinity as equivalent to the amount of strong acid (mol L^{-1}) required to lower the pH of a sample of (4.3). Several color indicators are available that have endpoints close to pH 4.3.

Selection of Potentiometric or Indicator Methods

Potentiometric titration directly measures the solutions ability to conduct electrical current through the use of a millivoltmeter or pH meter with suitable electrodes. Indicator methods rely on the color change of the solution to indicate that the desired endpoint has been reached.

A number of advantages make the potentiometric titrations the method of choice for accurate determinations. The equivalence point can be identified by the inflection in the titration curve or by the differential method of calculation (see "Gran Titration" section). The plot of a potentiometric titration curve also reveals any shift in the equivalence point caused by temperature, ionic strength, and, in the case of the total alkalinity, the effect of carbon dioxide concentration at the equivalence point. Moreover, the potentiometric method is free from the influence of color and turbidity, and individual idiosyncracies. Properly performed, however, the more rapid and simple indicator method is satisfactory for routine applications.

Equivalence Points

The equivalence point to which the total-alkalinity titration must be carried is determined by the concentration of CO_2 present. If the sample originally contained relatively little CO_2, and if the mixing during titration is not strong (reaeration), then the alkalinity will determine the equivalence point. A pH value of 4.8 is suggested as the equivalence point for the corresponding alkalinity concentration. When a great deal of CO_2 is present, or a high concentration of calcium carbonate exists in the sample, the equivalence point may vary slightly. Indicators effective in this range will give the most reliable results. A mixed indicator prepared from bromcresol green and methyl red is suitable for slightly higher equivalence points, where methyl orange can be used for those below 4.6. Since a written description of color shadings can frequently be misleading, buffer solutions of the applicable pH, with the proper volume of indicator added, should be prepared and used as standards for color comparison. The same considerations apply to the phenolophthalein end-point. For purposes of ascertaining the faint phenolphthalein end-point, it is advisable to employ a buffer solution with pH value of 8.3, to which proper volume of indicator is added, as a standard of comparison.

Sampling and Storage

For best results, samples should be collected in polyethylene or pyrex bottles. Because of the instability of samples containing considerable carbon dioxide, the alkalinity determinations should be performed as soon as practicable, and preferably within one day.

Procedure

PHENOLPHTHALEIN ALKALINITY

Add 0.1 mL (2 drops) of phenolphthalein indicator to a sample of suitable size, 50 or 100 mL if possible, in an Erlenmeyer flask. Titrate over a white surface with 0.02 N standard acid until the solution turns from pink to clear, indicating the corresponding equivalence point pH 8.3.

BROMCRESOL GREEN–METHYL RED

Total alkalinity by mixed bromcresol green–methyl red indicator method. Add 0.15 mL (3 drops) indicator to the solution in which the phenolphthalein alkalinity has been determined, or to a sample of suitable size, 50 or 100 mL if possible, in an Erlenmeyer flask. Titrate over a white surface with 0.02 N standard acid to the proper equivalence point. The indicator yields the following color responses:

> pH 5.2	greenish blue
pH 5.0	light blue with lavender–gray
pH 4.8	light pink–gray with a bluish cast
pH 4.6	light pink

METHYL ORANGE

Total alkalinity by methyl orange indicator method. Add 0.1 mL (2 drops) of indicator to the solution in which the phenolphthalein alkalinity has been determined, or to a sample of suitable size, 50 or 100 mL, in an Erlenmeyer flask. Titrate over a white surface with 0.02 N standard acid to the proper equivalence point. The indicator changes to orange at pH 4.6.

POTENTIOMETRIC

For greatest accuracy, titrate low alkalinities (<20 mg L^{-1} $CaCO_3$) potentiometrically rather than by indicator methods. Potentiometric titration avoids the error due to the sliding endpoint caused by free CO_2 in the sample at completion of the titration. With a microburette, titrate carefully a sample of suitable size, 100–200 mL, and record the volume C (in mL) of standard acid titrant (normality N) required to reach a pH of 4.5. Continue the titration to pH 4.2 and record the total volume D (in mL) of acid titrant.

GRAN TITRATION

Waters of low alkalinity (<100 μeq L^{-1}) may require a Gran titration since the equivalence pH is often higher than 4.3 and conventional titration tends to overestimate the alkalinity. The Gran titration uses a microburette and the endpoint inflection determined by plotting the change in pH per milliliter of acid added over pH. Once the equivalence pH is determined, the milliliters of acid added to that point are used to calculate solution alkalinity. The reader is referred to EPA (Environmental Protection Agency) methodologies developed from the acid-rain studies for calculation specifics.

CALCULATION

Indicator Methods:

$$\text{Phenolphthalein alkalinity as mg L}^{-1} \text{ CaCO}_3 = \frac{A \times N \times 5000}{\text{mL sample}}$$

$$\text{Total alkalinity as mg L}^{-1} \text{ CaCO}_3 = \frac{B \times N \times 5000}{\text{mL sample}}$$

Potentiometric method for low alkalinity:

$$\text{Total alkalinity as mg L}^{-1} \text{ CO}_3 = \frac{(2C - D)N \times 50{,}000}{\text{mL sample}}$$

where A = mL titration for sample to reach the phenolphthalein endpoint, B = total mL titration for sample to reach the second endpoint, C = mL titration for sample to reach pH 4.5, D = total mL titration for sample to reach pH 4.2, and N = normality of acid.

TYPICAL VALUES

	Alkalinity $(mg\ L^{-1}$ as $CaCO_3)$
Cache la Poudre River at canyon mouth near Fort Collins, Colorado	43.2
Tualatin River at West Linn, Oregon	34.5
Rogue River near Agness, Oregon	42.4
Colorado River at Stateline, Colorado	128
Colorado River at Imperial, California	147
Skagit River at Marblemount, Washington	21.4
Yellowstone River near Livingstone, Montana	73.8
Arkansas River near Coolidge, Kansas	210
Cumberland River near Grand River, Kentucky	52
Ogeechee River near Eden, Georgia	19
North Fork Whitewater River near Elba, Minnesota	258

Review

KEYWORDS

alkalinity   phenolphthalein
bicarbonate   endpoint
carbonate   titration
hydroxide   potentiometric
calcium carbonate   equivalence points

STUDY QUESTIONS

1. Calculate the average phenolphthalein alkalinity of your samples. Calculate the average total alkalinity of your sample for each method separately. Compare the results of the two methods. Which method produced more consistent results? Does this mean that this method is more accurate?
2. A solution turns pink immediately on the addition of methyl orange indicator (no titration). What does this mean about the alkalinity of the solution?
3. What problems might be encountered in measuring the alkalinity of water from a rapidly moving stream? Will field measured alkalinity be the same as laboratory measured alkalinity? Explain your answer.
4. Convert 1.0 mg L^{-1} of $CaCO_3$ alkalinity to units of meq L^{-1} of HCO_3^-

SAMPLE DATA SHEET

Analysis: Alkalinity   Instrument name:
Analyst: Instrument serial number:
Date: Calculations done by:
Start time: Calculations checked by:
 End: Comments:
Procedure:

SAMPLE IDENTIFICATION:	REPLICATE	MEASUREMENT
Standard	1	
	2	
Reported Value:	3	
	Mean:	
	Standard Deviation:	
	In Control?	

SAMPLE IDENTIFICATION:	REPLICATE	MEASUREMENT
	1	
	2	
Reported Value:	3	
	Mean:	
	Standard Deviation:	
	In Control?	

SAMPLE IDENTIFICATION:	REPLICATE	MEASUREMENT
	1	
	2	
Reported Value:	3	
	Mean:	
	Standard Deviation:	
	In Control?	

Suggested Readings

American Public Health Association, 1985. "Standard Methods for the Examination of Water and Wastewaters." Published by the Amer. Public Health Assoc., Amer. Water Works Assoc., and Water Poll. Control Fed., 16th Ed.

Bricker, O.P., and K.C. Rice. 1989. Acidic deposition to streams. *Environ. Sci. Technol.* **23(4)**: 379–385.

Hem, J.D. 1985. *Study and interpretation of the chemical characteristics of natural water.* United States Geological Survey Water Supply Paper 2254. Washington, D.C. 263 pp.

Nodvin, S.C., L.B. Weeks, E.P.E. Thomas, and L.J. Lund, 1986. Alkalization of a high-elevation Sierra Nevada Stream. *Water Res. Res.* **22(7)**: 1077–1082.

Omernick, J.M., and C.F. Powers, 1982. *Total alkalinity of surface waters: a national map.* U.S. Environmental Protection Agency. Govt. Printing Office. EPA–600/D–82–333.

Chapter 13

Hardness

Purpose: To introduce the concept of complex formation and stability, and to illustrate the analytical application of these concepts to the measurement of calcium + magnesium concentrations (hardness) in water.

"Hardness" is defined as a characteristic of water that represents the total concentration of calcium and magnesium ions expressed as calcium carbonate. If the hardness is numerically greater than the sum of the carbonate + bicarbonate alkalinity, the amount of hardness that is equivalent to the total alkalinity is called "carbonate hardness." The amount of hardness in excess of this is called "noncarbonate hardness." Originally the hardness of a water was understood to be a measure of the capacity of water to precipitate soap. Soap is precipitated chiefly by the calcium and magnesium ions commonly present in water, but it may also be precipitated by ions of other polyvalent metals such as aluminum, iron, manganese, strontium, and zinc, and also by hydrogen ions.

Calcium is dissolved from almost all rocks and soils, but the greatest concentrations are usually found in waters that have been in contact with limestone, dolomite, and gypsum. Most waters associated with granite or siliceous sand contain less than 10 mg L^{-1} of calcium; waters in areas where rocks are composed of dolomite and limestone contain 30–100 mg L^{-1}, and waters that have come in contact with deposits of gypsum may contain several hundred milligrams per liter (Hem, 1985).

Magnesium is dissolved from many rocks, especially from dolomitic rocks. The magnesium in soft water may amount to only 1 or 2 mg L^{-1} but water in areas that contain large quantities of dolomite or other magnesium-bearing rocks may contain 20 mg L^{-1} to several hundred milligrams per liter of magnesium.

Hardness is most frequently measured through application of the principle of chelation. Most metal ions are capable of sharing electron pairs with a donor—a species that has a "free" electron pair—to form a coordination bond. If a molecule or ion has more than one "free" electron pair that can be shared with a metal ion or similar species, it is called a "chelating agent." The complex is called a "chelate." Its stability is related to the number of coordination bonds that can be formed between the chelating agent and the metal ion.

One of the more commonly used chelating agents used in analytical chemistry is used for hardness determinations. It is ethylene diamine N,N,N′,N′-tetraacetic acid, usually abbreviated as EDTA. EDTA is a tetraprotic acid (p$K_{a,1}$ = 2.0, pH$_{a,2}$ = 2.8, pH$_{a,3}$ = 6.2, p$K_{a,4}$ = 10.3). In the completely deprotonated form, EDTA can form coordination bonds at six sites—the four oxygen and the two nitrogen sites. Most metal ions that have a coordination number of 6 form very stable complexes with the completely deprotonated EDTA. If the deprotonation of EDTA is not complete, these chelates are not as stable (Jenkins *et al.*, 1980).

In the determination of hardness with EDTA, several competing equilibria are involved. The sample solution is buffered at pH 10.0 as a compromise between chelate stability (EDTA chelate stability increases with increasing pH) and the need to prevent precipitation of the metal ions [e.g., as $CACO_3$(s) and $Mg(OH)_2$(s)] being analyzed. The ammonia buffer used in the test helps to prevent precipitation of metal ions since ammonia forms weak complexes with them. Since EDTA and its hardness complexes are not colored, and additional chelating agent, Eriochrome Black T(EBT), is used to facilitate endpoint detection. EBT is triprotic and exists primarily as the blue-colored divalent anion, HIn^{2+}, at pH 10. A small amount of EBT is added to the test solution prior to titration with EDTA and a red-colored complex is formed with Mg^{2+}:

$$HIn^{2-} \text{ (blue)} + Mg^{2+} \rightarrow H^+ + MgIn^- \text{ (red)} \qquad (1)$$

The buffer is frequently spiked with a trace of EDTA Mg^{2+} to facilitate endpoint detection in the unlikely event that the sample does not contain Mg^{2+}.

As EDTA is added to solution, it combines first with Ca^{2+}, and then with Mg^{2+} because the EDTA Ca^{2+} complex is more stable than the EDTA Mg^{2+} complex:

$$EDTA + Ca^{2+} \rightarrow EDTA\ Ca^{2+} \qquad K = 10^{+10.7} \qquad (2)$$

$$EDTA + Mg^{2+} \rightarrow EDTA\ Mg^{2+} \qquad K = 10^{+8.7} \qquad (3)$$

The metal ions are readily removed from their ammonia complexes because the EDTA–metal ion chelate is much more stable. After EDTA has complexed all of the free Mg^{2+} it will remove Mg^{2+} from the red EBT Mg^{2+} complex, causing reaction (1) to proceed from right to left and form the blue color of HIn^{2-}:

$$EDTA + MgIn^- \text{ (red)} \rightarrow EDTA\ Mg^{2+} \text{ (blue)} + HIn^{2-}$$

Procedure

There are two methods available for hardness determination. The first, called "hardness by calculation," involves performing a complete mineral analysis, then changing all concentrations and summing them. This is a very acurate, although time-consuming method. The EDTA titrimetric method, which measures only Ca^{2+} and Mg^{2+}, may have interferences associated with it. However, procedures are available for minimizing interference (APHA et al., 1985). Also, hardness contributed by other polyvalent ions is usually negligible.

EDTA TITRATION

Select a sample volume that requires less than 15 mL EDTA titrant. Do not extend duration of titration beyond 5 minutes, starting from time of buffer addition.

Dilute 25.0 mL of sample to about 50 mL with distilled water in a 150-mL beaker. Add 1–2 mL of buffer solution. Usually 1 mL will be sufficient to give a pH of 10.0–10.1. The absence of sharp endpoint color change in the titration usually means that an inhibitor must be added at this point in the procedure or that the indicator has deteriorated. Add one to two drops of indicator solution. Add the standard EDTA titrant slowly, with continuous stirring, with the last few drops at 3–5-second intervals. The color of the solution at the endpoint is blue under normal conditions. Daylight or a daylight fluorescent lamp is recommended; ordinary incandescent lights tend to produce a reddish tinge in the blue at the endpoint. If sufficient sample is available and interference is absent, accuracy may be improved by increasing the sample size.

Low-Hardness Sample

For natural waters of low hardness (<5 mgL^{-1}), take a larger sample, 100–1000 mL, for titration and add proportionately larger amounts of buffer, inhibitor, and indicator. Add the standard EDTA titrant slowly from a microburette and run a blank, using redistilled, distilled, or deionized (DI) water of the same volume as the sample, to which identical amounts of buffer, inhibitor, and indicator have been added.

Calculation

$$\text{Hardness (EDTA)} \atop \text{as mg L}^{-1} \text{ CaCO}_3 = \frac{A \times B \times 1000}{\text{mL sample}}$$

where A = mL titrant for sample and B = mg CaCO$_3$ equivalent to 1.00 mL EDTA titrant.

REPORTING OF RESULTS

When reporting hardness, state either the ions determined or the method used, for example, "hardness (Ca, Mg)," "hardness (Ca, Mg, Sr, Be, Al, etc.)," "hardness (EDTA)."

TYPICAL VALUES

	Hardness (mg L^{-1} as $CaCO_3$)
Cache la Poudre River at canyon mouth near Fort Collins, Colorado	43
Tualatin River at West Linn, Oregon	45
Rogue River near Agness, Oregon	39
Colorado River at Stateline, Colorado	330
Colorado River at Imperial, California	350
Skagit River at Marblemount, Washington	22
Yellowstone River near Livingstone, Montana	75
Arkansas River near Coolidge, Kansas	1425
Cumberland River near Grand River, Kentucky	128
Ogeechee River near Eden, Georgia	33
North Fork Whitewater River near Elba, Minnesota	437

Review

KEYWORDS

hardness
chelation
EDTA

STUDY QUESTIONS

1. What is meant by chelation? Give an example.
2. Name two polyvalent ions, other than calcium and magnesium, that may be responsible for hardness in water.
3. How do chemical water softeners work?
4. What is the advantage of endpoint enhancement?
5. Determine the hardness of a 50-mL water sample, and then repeat the determination using 100 mL of the same sample. Discuss any differences.

SAMPLE DATA SHEET

Analysis: Hardness   Instrument name:

Analyst: Instrument serial number:

Date: Calculations done by:

Start time: Calculations checked by:

 End: Comments:

Procedure:

SAMPLE IDENTIFICATION:	REPLICATE	MEASUREMENT
Standard	1	
	2	
Reported Value:	3	
	Mean:	
	Standard Deviation:	
	In Control?	

SAMPLE IDENTIFICATION:	REPLICATE	MEASUREMENT
	1	
	2	
Reported Value:	3	
	Mean:	
	Standard Deviation:	
	In Control?	

SAMPLE IDENTIFICATION:	REPLICATE	MEASUREMENT
	1	
	2	
Reported Value:	3	
	Mean:	
	Standard Deviation:	
	In Control?	

Suggested Readings

American Public Health Association, 1985. "Standard Methods for the Examination of Water and Wastewaters." Published by the Amer. Public Health Assoc., Amer. Water Works Assoc., and Water Poll. Control Fed., 16th Ed.

Hem, J.D., 1985. *Study and interpretation of the chemical characteristics of natural water.* U.S. Geological Survey Water Supply Paper 2254 Washington. D.C. 263 pp.

Jenkins, D., V.L. Snoeyink, J.F. Ferguson, and J.O. Leckie, 1980. "Water chemistry: laboratory manual." Third Edition. 183 pp.

Sawyer, C.N., and P.L. McCarty. 1978. "Chemistry for Environmental Engineering." McGraw-Hill, New York. 532 pp.

Chapter 14

Chloride

Purpose: To illustrate the concept of complex formation and its application to the quantitative analysis of chloride in water.

Chloride is present in all natural waters, but concentrations are generally low, less than 1 mg L^{-1}. Exceptions occur where streams receive inflows of high chloride groundwater or industrial waste or are affected by oceanic influences. The major sources of chloride appear to be igneous rocks, liquid inclusions, sedimentary rocks, recycled chloride from the oceans, and volcanic gases and hot springs. Feldspathoid sodalite and apatite are among the igneous minerals that contain chloride in fairly low concentrations. Chloride may replace hydroxide in biotite and hornblende, be present in liquid inclusions, or be in solution in glassy rocks. Igneous rocks cannot yield very high concentrations of chloride to normally circulating natural water (Hem, 1985).

A more important source of chloride is the sedimentary rocks, particularly the evaporites. When porous rocks are submerged by the sea at any time after their formation, they become impregnated with soluble salts, in which chloride plays a major role. The chloride of marine evaporite sediments is recycled when it goes into solution in streamflow (Hem, 1985).

A relatively large amount of chloride is recycled from the oceans by precipitation. As would be expected, rainwater close to the ocean is more concentrated than that farther inland. Near the coast the rainwater commonly contains 1 mg L^{-1} or more, but decreases rapidly to an average of 0.2 mg L^{-1} inland.

The chemical behaviour of chloride in natural water is conservative. Chloride ions do not significantly enter into oxidation or reduction reactions, form no important solute complexes with other ions, do not form salts of low solubility, are not significantly absorbed on mineral surfaces, and play few vital biochemical roles. The circulation of chloride ions in the hydrologic cycle is largely through physical processes and may be used to calculate the water balance.

Chloride ions move with water through most soils and loosely compacted rock. Because the chloride ion is physically large compared to many of the other major ions in water, it could be expected to be held back in interstitial or pore water in clay and shale while water itself was transmitted. Chloride anion sorption by soils is minimal.

The most common type of water in which chloride is the dominant anion is one in which sodium is the predominant cation. This commonly happens in rainfall near the ocean, brines near saturation with respect to sodium chloride, seawater, and a few acid waters in which sodium is not the dominant cation. Chlorides exist in natural waters mostly as metallic salts $MgCl_2$, $NaCl$, or $CaCl_2$ (Hem, 1985).

Although high concentrations of Cl in water have not been found to be toxic to humans, chloride may severely corrode metal pipes and kill many types of plants. You can "taste" a concentration of 250 mg L^{-1} Cl^- when NaCl is present, but up to 1000 mg L^{-1} Cl^- can be present in form of $MgCl_2$ or $CaCl_2$ before it tastes salty. It is this "taste effect" that is responsible for the maximum allowable limit of 250 mg L^{-1} Cl^- in drinking water. The concentrating process of chloride is mainly through evaporation.

The "Argenometric Method" (see below) for chloride determination involves the titration of a chloride containing sample with silver nitrate ($AgNO_3$). A potassium chromate (K_2CrO_4) indicator is used to detect the endpoint. When $[Cl^-]$ concentration is greater than $[Ag^+]$ concentration, silver chloride (AgCl) will precipitate. However, when $[Ag^+]$ concentration exceeds $[Cl^-]$ concentration, Ag^+ will combine with the chromate ion (CrO_4^-) of the indicator to form red silver chromate. Thus the endpoint of chloride titration is a change to a pinkish yellow solution color. Silver chloride is precipitated quantitatively before red silver chromate is formed. The stoichiometric relations are

$$AgNO_3 + Cl^- \rightarrow AgCl + NO_3^-$$

$$2Ag^+ + K_2CrO_4 \text{ (yellow)} \rightarrow Ag_2CrO_4 \text{ (red)} + 2K^+$$

Samples must be adjusted to the pH range 7–10 before titrating.

Possible interference with chloride determinations are caused by bromide and iodide registering as equal amounts of chloride or chromate, ferric, and sulfide ions in excess of 10 mg L^{-1} may mask the end point (APHA et al., 1985).

Procedure

ARGENOMETRIC METHOD

Apparatus

1. Graduated cylinder (at least 100-mL capacity).
2. Several 250-mL beakers.
3. Magnetic stirrer and spinbars.
4. Burette filled with $AgNO_3$.
5. 1 mL automatic pipettor.

Procedure

Using a graduated cylinder, measure a 100-mL sample and pour it into a 250-mL beaker.

1. A 100 mL-sample is recommended, but for sample with a very high Cl concentration, a smaller amount should be used. If you use less than 100 mL, however, you must dilute the sample to 100 mL with DI water. In general, start with 100 mL of sample. If more than 25 mL of $AgNO_3$ is needed for titration, throw the sample out and try 50 mL, then 25 mL, then 10. Probably no dilutions will be necessary.
2. Check the pH of the sample. If it is between 7 and 10, go to step 3. If it isn't, use either 1 N NaOH (to raise it) or 1 N H_2SO_4 (to lower it).
3. Add 1 mL K_2CrO_4 indicator solution. This will turn your sample a bright yellow. Place a spinbar in the beaker (don't splash)

and place the beaker on the stirrer. (It is very important to keep the sample well mixed while titrating). With a full burette and the spinbar spinnning slowly, start titrating. After adding titrant always pause and allow the reaction to go to completion. You will notice a white precipitate forming (AgCl), and this will cloud the sample. When you start seeing a bit of red $AgCrO_4$ just as the titrant enters the sample, start adding titrant drop by drop until the sample turns a pinkish cloudly yellow. Endpoint determination is left up to the individual, so be consistent when you run other samples or trials. Do sample replicates.

4. Now that you've had some practice, run a blank (100 mL of DI water).

5. To calculate the mg L^{-1} Cl:

$$\text{mgL}^{-1}\ Cl = \frac{(A-B) \times N \times 35450}{\text{mLs of sample}}$$

where A = mL titrant for sample, B = mL titrant for blank, and N = normality of $AgNO_3$ (0.0141).

TYPICAL VALUES	
	Chloride (mg L^{-1})
Cache la Poudre River at canyon mouth near Ft. Collins, Colorado	1.94
Tualatin River at West Linn, Oregon	9.41
Rogue River near Agness, Oregon	2.54
Colorado River at Stateline, Colorado	66.1
Colorado River at Imperial, California	112
Skagit River at Marblemount, Washington	0.12
Yellowstone River near Livingstone, Montana	7.91

TYPICAL VALUES (Continued)	
Arkansas River near Coolidge, Kansas	140
Cumberland River near Grand River, Kentucky	3.7
Ogeechee River near Eden, Georgia	5.2
North Fork Whitewater River near Elba, Minnesota	5.5

Review

KEYWORDS

argentometric
chloride
ocean

STUDY QUESTIONS

1. Compute the average chloride concentration of the samples. Show your work.
2. Explain, in your own words, how the addition of silver nitrate to a water sample yields information about the chloride content of the sample.
3. Given that white silver chloride precipitates before red silver chromate is formed, which is more soluble, $AgCl$ or Ag_2CrO_4? Explain your answer.
4. Cl^- passes unchanged through the digestive system of humans and other mammals, and is a micronutrient for plants, required in varied quantities. What kind of spatial and temporal fluctuations in Cl^- concentration would you expect due to biotic influences in lakes and streams?

SAMPLE DATA SHEET

Analysis: Chloride   Instrument name:

Analyst: Instrument serial number:

Date: Calculations done by:

Start time: Calculations checked by:

  End: Comments:

Procedure:

SAMPLE IDENTIFICATION:	REPLICATE	MEASUREMENT
Standard	1	
	2	
Reported Value:	3	
	Mean:	
	Standard Deviation:	
	In Control?	

SAMPLE IDENTIFICATION:	REPLICATE	MEASUREMENT
	1	
	2	
Reported Value:	3	
	Mean:	
	Standard Deviation:	
	In Control?	

SAMPLE IDENTIFICATION:	REPLICATE	MEASUREMENT
	1	
	2	
Reported Value:	3	
	Mean:	
	Standard Deviation:	
	In Control?	

Suggested Readings

American Public Health Association, 1985. "Standard Methods for the Examination of Water and Wastewaters." Published by the Amer. Public Health Assoc., Amer. Water Works Assoc., and Water Poll. Control Fed., 16th Ed. pp. 286–294.

Hem, J.D., 1985. *Study and interpretation of the chemical characteristics of natural water.* U.S. Geological Survey Water Supply Paper 2254. Washington, D.C. 263 pp.

Chapter 15

Sulfate

Purpose: To determine the sulfate (SO_4^{2-}) concentration in a water sample, and to understand the principles of solution precipitation.

Sulfur is not a major constituent of the earth's crust, but occurs as sulfides in igneous and sedimentary rocks. The sulfides, when in contact with air and water, oxidize to form sulfate (SO_4^{2-}) and hydrogen ions. The metal sulfide oxidation may result in acid mine drainage, waters of low pH with increased concentrations of mobile metal ions. Sulfate is common in evaporite sediments, particularly as gypsum ($CaSO_4$).

Sulfate is chemically stable in aerated water, but may form ion pairs and metal complexes, and is involved with biologic processes since it is a required macronutrient for plant life. Sulfate concentrations in precipitation are variable, but surprisingly high. The sulfate in rainfall may be from hydrogen sulfide (H_2S) emissions from the ocean, certain anaerobic waters, or other sulfur emissions from volcanoes, fumaroles, or hot springs. The effect of local air pollution, and even global cycling of sulfur compounds may contribute to the sulfate content in precipitation.

Sulfur may be found in living matter, principally as a component of amino acids in proteins. The sulfur cycle is similar to the nitrogen cycle in that gaseous and hydrologic transfers may be involved. Salts of calcium and magnesium sulfate are often associated with alkali soils. Sulfate concentrations may range from 1 to over 1000 mg L^{-1}.

There are two variations of the barium chloride ($BaCl_2$) precipitation method for the determination of SO_4^{2-} Sulfate is precipitated in a hydrochloric acid (HCl) solution as barium sulfate ($BaSO_4$) by the addition of barium chloride ($BaCl_2$). The precipitation is carried out near the boiling temperature, and after a digestion period, the precipitate is filtered, washed, and weighed as $BaSO_4$. The second procedure is to precipitate sulfate in an acetic acid medium with barium chloride ($BaCl_2$) to form barium sulfate ($BaSO_4$) crystals. Light absorbance of the $BaSO_4$ suspension is measured by a photometer and the SO_4^{2-} concentration is determined from a standard curve.

Procedure

HACH TURBIDIMETRIC METHOD

Apparatus

1. Spectrophotometer, set at 450 nm.
2. Several 50-mL Erlenmeyer flasks, or sample cells.
3. Several rubber stoppers.
4. Graduated cylinder.

Procedure

1. Measure a 25-mL sample into a 50-mL Erlenmeyer flask.
2. Add the contents of one SulfaVer 4 Sulfate Reagent Powder Pillow and swirl to mix. A white turbidity will develop if sulfate is present. Allow at least 5 minutes for the turbidity to develop fully, but complete analysis within 10 minutes.
3. Adjust the spectrophotometer by zeroing with untreated original water sample.
4. To find the milligrams per liter sulfate, compare your results to a standard curve.

TYPICAL VALUES

	Sulfate (SO_4^{2-}) (mg L^{-L})
Cache la Poudre River at canyon mouth near Ft. Collins, Colorado	5.86
Tualatin River at West Linn, Oregon	9.95
Rogue River near Agness, Oregon	3.47
Colorado River at Stateline, Colorado	240
Colorado River at Imperial, California	309
Skagit River at Marblemount, Washington	3.73
Yellowstone River near Livingstone, Montana	26.6
Arkansas River near Coolidge, Kansas	1925
Cumberland River near Grand River, Kentucky	15
Ogeechee River near Eden, Georgia	4.7
North Fork Whitewater River near Elba, Minnesota	13

Review

KEYWORDS

sulfate	precipitate
sulfur	turbidimetric (sulfate)

STUDY QUESTIONS

1. Convert 4.8 mg L^{-1} of SO_4^{2-} to mg L^{-1} of SO_4–S.
2. Define a system where SO_4^{2-} is the dominant mobile anion.
3. The solubility product of gypsum is $10^{-4.5}$. Calculate the concentration of Ca^{2+} at dissolution equilibrium in meq L^{-1}.
4. What chemical reactions are involved in the oxidation of pyrite as they relate to acid mine drainage?
5. A subalpine stream in the Colorado Rocky Mountains has a mean annual concentration of 6.2 meq L^{-1} of SO_4^{2-} and an annual water yield of 75 cm. Calculate the annual flux in equivalents per hectare per year (eq ha^{-1} yr^{-1}).

SAMPLE DATA SHEET

Analysis: Sulfate   Instrument name:

Analyst: Instrument serial number:

Date: Calculations done by:

Start time: Calculations checked by:

  End: Comments:

Procedure:

SAMPLE IDENTIFICATION:	REPLICATE	MEASUREMENT
Standard	1	
	2	
Reported Value:	3	
	Mean:	
	Standard Deviation:	
	In Control?	

SAMPLE IDENTIFICATION:	REPLICATE	MEASUREMENT
	1	
	2	
Reported Value:	3	
	Mean:	
	Standard Deviation:	
	In Control?	

SAMPLE IDENTIFICATION:	REPLICATE	MEASUREMENT
	1	
	2	
Reported Value:	3	
	Mean:	
	Standard Deviation:	
	In Control?	

Suggested Readings

American Public Health Association, 1985. "Standard Methods for the Examination of Water and Wastewaters." Published by the Amer. Public Health Assoc., Amer. Water Works Assoc., and Water Poll. Control Fed., 16th Ed. pp. 464–468.

Hem, J.D., 1985. *Studys and interpretation of the chemical characteristics of natural waters.* U.S. Geological Survey Water Supply Paper 1473. Washington. D.C. 363 pp.

Hach Chemical Company, 1984. "Chemical Analysis of Water." pp. (2–277)–(2–278).

Chapter 16

Dissolved Oxygen

Purpose: To learn ways to measure dissolved oxygen (DO) in water and to understand some of the natural processes that affect DO.

All living organisms are dependent on oxygen in one form or another to maintain the metabolic processes that produce energy for growth and reproduction. Aerobic processes are the subject of greatest interest because of their need for free oxygen.

Temperature has the greatest affect on the solubility of oxygen in water. The relationship between oxygen solubility and temperature is calculated using an empirical relationship (Churchill et al., 1957):

$$C_{s(t)} = 14.652 - 0.41022T + 0.0079917T^2 - 0.000077774T^3$$

where $C_{s(t)}$ is the solubility of oxygen in water at a given temperature, T (°C). The relationship shows that temperature is inversely proportional to solubility, meaning that as water becomes warmer its ability to hold oxygen decreases.

Because rates of biological oxidation increase with temperature, and oxygen demand increases accordingly, high-temperature conditions where dissolved oxygen is least soluble, are of greatest concern to land managers. Most of the critical conditions related to dissolved-oxygen deficiency occur during the summer months when temperatures are high and solubility of oxygen is at a minimum.

All the atmospheric gases are soluble in water to some degree. Both nitrogen and oxygen are classed as poorly soluble, and since they do not react with water chemically, their solubility is directly proportional to their partial pressures. Henry's law may be used to calculate the amounts present at saturation at any given temperature.

The solubility of atmospheric oxygen in fresh waters ranges from 14.6 mg L^{-1} at 0°C to about 7 mg L^{-1} at 35°C under 1 atm of pressure. Since it is a poorly soluble gas, its solubility varies directly with the atmospheric pressure at any given temperature. This is an important consideration at higher altitudes.

The solubility of oxygen in water at a given temperature as a function of pressure is calculated using the formula

$$C_{s(p)} = C_s \frac{P - p}{760 - p}$$

where $C_{s(p)}$ is the solubility of oxygen in water (mg L^{-1}) at a given temperature and pressure, C_s is the solubility (mg L^{-1}) at the same temperature but standard pressure (760 mm Hg), P is barometric pressure (mm Hg), and p is the pressure of saturated water vapor at the given temperature. The effects of temperature and pressure on oxygen solubility can be modeled by combining these two equations. The result is a family of curves, which yield a unique value of oxygen solubility for each combination of temperature and pressure (Figure 16.1).

Because of its involvement in so many chemical and biological reactions and its sensitivity to physical conditions, dissolved oxygen is one of the least conservative constituents of water quality. In warm waters with very high productivity, diel variations of 10 mg L^{-1} may be observed. Spatial and temporal variability should be considered when collecting and interpreting data (Hem, 1985).

The ultimate source of oxygen in water exposed to air is the atmosphere, but some oxygen is contributed by an indirect route as a by-product of photosynthesis. Waterbodies in which there is much organic productivity often display wide fluctuations of dissolved oxygen in response to the biological activity. The relative importance of organic sources and atmospheric ones perhaps can be evaluated by comparing productivity with the hydraulic parameters that essentially control absorption of oxygen from the air at a water surface.

Owing to the rapidly changing input and consumption rates, the oxygen content of a surface-water body or stream is a highly

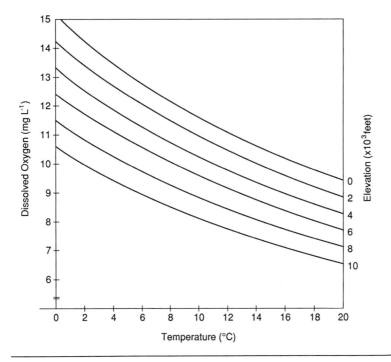

Figure 16.1 Dissolved oxygen (D.O.) saturation decreases with increased water temperature and increased elevation (partial pressure).

transient property. A measurement is meaningful only for the spot of sampling and a brief time period. The oxygen content of a sample of water can readily change after collection and thus must be chemically preserved and determined quickly. The development of electrodes for sensing dissolved oxygen has greatly simplified sampling and determination problems.

Collection of Samples

A certain amount of care must be exercised in the collection of samples to be used for dissolved-oxygen determinations. In most cases of interest, the dissolved-oxygen level will be below saturation, and exposure to the air may lead to erroneous results. For this reason, dissolved oxygen samplers are designed on the principle that contact with air cannot be avoided during the time the sample bottles are being filled. If space is available to allow the bottles to overflow, a sample of water that is representative of the mixture being sampled can be obtained. Most samplers are designed to provide an overflow of two or three times the bottle volume to ensure collection of representative samples.

Most samples for dissolved oxygen are collected in the field, where it is not convenient to perform the entire determination. Since oxygen values may change radically with time because of biological activity, it is customary to "fix" the samples immediately after collection. The usual procedure is to treat the samples with the conventional reagents used in the dissolved-oxygen test and then perform the titration when the samples are brought to the laboratory. Better results are also obtained if the fixed samples are stored in the dark and on ice until the analyses can be completed. The chemical treatment employed in fixing is radical enough to arrest all biological action, and the final titration may be delayed for up to 6 hours.

Determination Of Dissolved-Oxygen

THE WINKLER METHOD

The Winkler method and its modifications are the standard procedures for determining dissolved oxygen. The test depends on the fact that oxygen oxidizes Mn^{2+} to a higher state of valence under alkaline conditions and that manganese in high states of valence is capable of oxidizing I^- to free I_2 under acid conditions. Thus the amount of free iodine released is equivalent to the dissolved oxygen originally present. The iodine is measured with a standard sodium thiosulfate solution and interpreted in terms of dissolved oxygen.

The Winkler method is subject to interference from a great many substances. Certain oxidizing agents such as nitrite and Fe^{3+} are capable of oxidizing I^- to I_2 and produce results that are too high. Reducing agents such as Fe^{2+}, SO_3, S^-, and polythionates, reduce I_2 to I^- and produce results that are too low. The unmodified Winkler method is applicable only to relatively pure waters (APHA et al., 1985).

The reactions involved in the Winkler procedure are as follows:

$$Mn^{2+} + 2OH^- \leftrightharpoons Mn(OH)_2 \qquad \text{white precipitate}$$

If no oxygen is present, a pure white precipitate of Mn $(OH)_2$ forms when $MnSO_4$ and the alkali–iodide reagent $(NaOH + KI)$ are added to the sample. If oxygen is present in the sample, then some of the Mn^{2+} is oxidized to a higher valence and precipitates as a brown hydrated oxide. The reaction is usually represented as follows:

$$Mn^{2+} + 2OH^- + \tfrac{1}{2}O_2 \leftrightharpoons MnO_2 + H_2O$$

or

$$Mn(OH)_2 + \tfrac{1}{2}O_2 \leftrightharpoons MnO_2 + H_2O$$

The oxidation of Mn^{2+} to MnO_2, sometimes called "fixation of the oxygen," occurs slowly, particularly at low temperatures. Furthermore, it is necessary to move the flocculated material ("floc") throughout the solution to enable all the oxygen to react. Vigorous shaking of the samples for at least 20 seconds is needed.

After shaking the samples for a time sufficient to allow all oxygen to react, the floc is allowed to settle so as to leave at least 5 cm of clear liquid below the stopper; then sulfuric acid is added. Under the low pH conditions that result, the MnO_2 oxidized I^- to produce free I_2.

$$MnO_2 + 2I^- + 4H^+ \leftrightharpoons Mn^{2+} + I_2 + 2H_2O$$

The sample should be stoppered and shaken for at least 10 seconds to allow the reaction to go to completion and to distribute the iodine uniformly throughout the sample.

The sample is now ready for titration with $N/40$ thiosulfate. The use of $N/40$ thiosulfate is based on the premise that a 200-mL sample will be used for titration. In adding the reagents used for the Winkler test, a certain amount of dilution of the sample occurs; therefore, it is necessary to take a sample somewhat greater than 200 mL for the titration. When 300-mL bottles are used in the test, 2 mL of $MnSO_4$ and 2 mL of alkali–KI solutions are used. These are added in such a manner as to displace approximately 4 mL of sample from the bottle; and a correction should be made. When the 2 mL of acid is added, none of the oxidized floc is displaced; thus no correction need be made for its addition. To correct for the addition of the first two reagents, 203 mL of the treated sample is taken for titration.

Titration of a sample of a size equivalent to 200 mL of the original sample with $N/40$ thiosulfate solution yields results in milliliter values, which can be interpreted directly in terms of milligrams per liter of dissolved oxygen.

Azide Modification of the Winkler Method

The nitrite ion is one of the most frequent interferences encountered in the dissolved-oxygen determination. It occurs principally in effluents from sewage treatment plants employing biological processes, in river waters, and in incubated biochemical oxygen demand (BOD) samples. Nitrite interference may be easily overcome by the use of sodium azide (NaN_3). It is most convenient to incorporate the azide in the alkali–KI reagent. When sulfuric acid is added, the NO_2^- is destroyed. By this procedure, nitrite interference is eliminated, and the method of determination retains the simplicity of the original Winkler procedure.

Membrane Electrode Method

There are a number of advantages to using a dissolved-oxygen meter instead of titration methods. The meter with its probe can be used in situ, eliminating much of the error caused by sampling technique and storage and transfer of samples.

The oxygen probe has an electrode system covered by a plastic membrane that is permeable to oxygen but serves as a diffusion carrier for impurities. This helps to eliminate interference in highly colored or polluted waters. Under steady-state conditions the current is directly proportional to the DO concentration. Temperature and salinity will affect measurements—most meters provide adjustments for these.

Procedure

Azide Modification

1. To the sample as collected in a 250–300-mL bottle, add 2 mL of manganese sulfate solution, followed by 2 mL of alkali–iodide–azide reagent, well below the surface of the liquid; stopper with care to exclude air bubbles, and mix by inverting the bottle at least 15 times.
2. When the precipitate settles, leaving a clear supernate above the manganese hydroxide floc, and shake again.
3. After at least 2 minutes of settling has produced 100 mL of clear supernate, carefully remove the stopper and immediately add 2.0 mL of concentrated H_2SO_4 by allowing the acid to run

down the neck of the bottle, restopper, and mix by gentle inversion until dissolution is complete.

4. The iodine should be uniformly distributed before decanting the amount needed for titration. This should correspond to 200 mL of the original sample after correction for the loss of sample by displacement with the reagents has been made. Thus, when a total of 4 mL (2 mL each) of the manganese sulfate and alkali–iodide–azide reagents is added to a 300-mL bottle, the volume taken for titration should be $200 \times 300/(300 - 4) = 203$ mL.

5. Titrate with 0.0250 N thiosulfate solution to pale straw color. Add 1–2 mL starch solution and continue the titration to the first disappearance of the blue color. Because 1 mL 0.025 N sodium thiosulfate titrant is equivalent to 0.200 mg DO, each milliliter of sodium thiosulfate titrant used is equivalent to 1 mg L DO when a volume equal to 200 mL of original sample is titrated. If a volume of sample other than 200 mL is used, the equation is

$$\text{mg L}^{-1} = \frac{(\text{mL titrant})\,(\text{normality thiosulfate})\,(8000)}{\text{mL sample}\,(\text{mL of bottle} - 4)/\text{mL of bottle titrated}}$$

Membrane Electrode Method

Probe and membrane should be prepared as per manufacturer's instruction. Most electrode instruments operate in the following format.

1. Place probe in sample and agitate to assure circulation over membrane.
2. Turn switch to TEMP and read temperature. Set O_2 solubility factor dial to observed temperature, using proper salinity index.
3. Turn switch to O_2 and read DO value in milligrams per liter directly from meter dial.

Hach-Dissolved-Oxygen Test

1. Fill the glass-stoppered DO bottle with the water to be tested by allowing the water to overflow the bottle for 2 or 3 minutes. Be certain there are no air bubbles present in the bottle.

2. Add the contents of one pillow each of Dissolved Oxygen 1 Reagent Powder and Dissolved Oxygen 2 Reagent Powder. Stopper firmly and carefully so that no air is trapped in the bottle. Grip the bottle and shake vigorously to mix. A flocculent precipitate will form. If oxygen is present, the precipitate will be brownish-orange in color.

3. Allow the sample to stand until the floc has settled halfway and leaves the upper half of the bottle clear. Then again shake the bottle and again let it stand until the upper half of the bottle is clear.

4. Remove the stopper and add the contents of one pillow of Dissolved Oxygen 3 Reagent Powder. Carefully restopper and shake to mix. The floc will dissolve and a yellow color will develop if oxygen is present. This is the prepared sample.

5. Fill the plastic measuring tube level full with prepared sample and pour it into the mixing bottle.

6. While swirling the sample to mix, add PAO Titrant dropwise, counting each drop, until the sample changes from yellow to colorless. The dropper must be held in a vertical manner. Each drop is equal to 1 mg L^{-1} DO.

TYPICAL VALUES

	Dissolved Oxygen (mg L^{-1})
Cache la Poudre River at canyon mouth near Ft. Collins, Colorado	9.4
Tualatin River at West Linn, Oregon	9.3
Rogue River near Agness, Oregon	10.5
Colorado River at Stateline, Colorado	8.8
Colorado River at Imperial, California	8.7
Skagit River at Marblemount, Washington	12.1
Yellowstone River near Livingstone, Montana	9.6
Arkansas River near Coolidge, Kansas	9.8
Cumberland River near Grand River, Kentucky	8.7
Ogeechee River near Eden, Georgia	8.4
North Fork Whitewater River near Elba, Minnesota	7.2

Review

KEYWORDS

dissolved oxygen
oxygen
N/40 thiosulfate
Winkler
azide modification
membrane electrode

STUDY QUESTIONS

1. Calculate the saturated dissolved oxygen concentration for a surface water at 4°C at 635 mm Hg pressure. Calculate the DO for a sample at 4°C and 735 mm Hg pressure. Discuss the effect of elevation on DO concentrations.
2. How would you expect the DO concentrations of a shallow, productive lake to fluctuate throughout the course of a day? What about a rapidly moving stream? Consider both temperature and biotic effects.
3. How would you expect DO concentrations to vary in a large, deep lake from the shoreline to open water?
4. Discuss the effect of large organic matter on stream reaeration and the effect of fine organic matter on the biological oxygen demand (BOD). What can land managers do to maintain appropriate DO levels?

SAMPLE DATA SHEET

Analysis: Dissolved Oxygen   Instrument name:

Analyst: Instrument serial number:

Date: Calculations done by:

Start time: Calculations checked by:

 End: Comments:

Procedure:

SAMPLE IDENTIFICATION:	REPLICATE	MEASUREMENT
Standard	1	
	2	
Reported Value:	3	
	Mean:	
	Standard Deviation:	
	In Control?	

SAMPLE IDENTIFICATION:	REPLICATE	MEASUREMENT
	1	
	2	
Reported Value:	3	
	Mean:	
	Standard Deviation:	
	In Control?	

SAMPLE IDENTIFICATION:	REPLICATE	MEASUREMENT
	1	
	2	
Reported Value:	3	
	Mean:	
	Standard Deviation:	
	In Control?	

Suggested Readings

American Public Health Association, 1985. "Standard Methods for the examination of Water and Wastewater." Published by the Amer. Public Health Assoc., Amer. Water Works Assoc., and Water Poll. Control Fed., 16th Ed. pp. 413-426.

Berry, J.D., 1975. *Modeling the impact of logging debris on the dissolved oxygen balance of small mountain streams.* M.S. Thesis, Oregon State Univ. 163 pp.

Brazier, J.R., and G.W. Brown, 1973. *Buffer strips for stream temperature control.* Research Paper 15. Forest Res. Laboratory, Oregon State Univ. 9 pp.

Brown, G.W., 1983. "Forestry and water quality." (2nd Ed.). Oregon State University Book Stores, Inc., Corvallis, Oregon. 142 pp.

Brown, G.W., and J.T. Krygier, 1970. Effects of clearcutting on stream temperature. *Water Resources Research* **6(4)**: 1133-1139.

Brown, G.W., 1969. Predicting temperatures of small streams. *Water Resources Research* **5(1)**: 68-75.

Churchill, M.A., R.A. Buckingham, and H.K. Elmore, 1957. *The prediction of stream reaeration rates.* Tennessee Valley Authority, Division of Health and Safety, Chattanooga, Tennessee. 98 pp.

Coble, D.W., 1961. Influence of water exchange and dissolved oxygen in redds on survival of steelhead trout embryos. *Trans. Amer. Fish. Soc.* **90(4)**: 469-474.

Froehlich, H.A., 1973. Natural and man-caused slash in headwaters streams. *In* "Loggers Handbook," Vol. 33 pp. 15-17, 66-70, 82-86.

Hem, J.D. 1985. Study and interpretation of the chemical characteristics of natural water. U.S. Geological Survey. Water Supply Paper 2254. 263pp.

Chapter 17

Biochemical Oxygen Demand

Purpose: Familiarization with the concepts of biochemical oxygen demand (BOD).

Biochemical oxygen demand (BOD) is usually defined as the amount of oxygen required by bacteria while stabilizing decomposable organic matter under aerobic conditions. The term "decomposable" may be interpreted as meaning that the organic matter can serve as food for the bacteria, and energy is derived from its oxidation.

The BOD test is widely used to determine the strength of domestic and industrial wastes discharged into natural water courses in which aerobic conditions exist in terms of the oxygen required. Decomposition of organic matter will also require a certain oxygen level. This test is of prime importance in regulatory work and in studies designed to evaluate the purification capacity of receiving bodies of water (Sawyer and McCarty, 1978).

The BOD test is a bioassay procedure involving the measurement of oxygen consumed by living organisms (mainly bacteria) while utilizing the organic matter present in a water sample, under conditions as similar as possible to those that occur in nature. In order to make the test quantitative, the samples must be protected from the air to prevent reaeration as the dissolved-oxygen level diminishes. In addition, because of the limited solubility of oxygen in water, strong wastes or organic loads must be diluted to levels of demand in keeping with this value to ensure that dissolved oxygen will be present throughout the period of the test. Since this is a bioassay procedure, it is extremely important that environmental conditions be suitable for the living organisms to function in an unhindered manner at all times. This condition means that toxic substances must be absent and that all accessory nutrients needed for bacterial growth, such as nitrogen, phosphorus, and certain trace elements, must be present. Biological degradation of organic matter under natural conditions is brought about by a diverse group of organisms that carry the oxidation essentially to completion, that is, almost entirely to carbon dioxide and water. Therefore, it is important that a mixed group of organisms, commonly called "seed," be present in the test.

The BOD test may be considered as a wet oxidation procedure in which the living organisms serve as the medium for oxidation of the organic matter to carbon dioxide and water. A quantitative relationship exists between the amount of oxygen required to convert a definite amount of any given organic compound to carbon dioxide, water, and ammonia. On the basis of this relationship, it is possible to interpret BOD data in terms of organic matter, as well as the amount of oxygen used.

The oxidative reactions involved in the BOD test are a result of biological activity, and the rate at which the reactions proceed is governed to a major extent by population numbers and temperature. Temperature effects are held constant by performing the test at 20°C, which is, more or less, a median value as far as natural bodies of water are concerned. The predominant organisms responsible for the stabilization of organic matter in natural waters are forms native to the soil. The rate of their metabolic processes at 20°C and under the conditions of the test is such that time must be reckoned in days. Theoretically, an infinite time is required for complete biological oxidation of organic matter, but for all practical purposes, the reaction may be considered complete in 20 days. However, a 20-day period is too long to wait for results in most instances. It has been found by experience that a reasonably large percentage of the total BOD is exerted in 5 days; consequently the test has been developed on the basis of a 5-day incubation period. It should be remembered, therefore, that 5-day BOD values represent only a portion of the total BOD. In the case of domestic and many industrial wastewaters, it has been found that the 5-day BOD value is about 70 to 80 percent of the total BOD (Sawyer and McCarty, 1978).

Nitrogen-containing organic compounds can be degraded in the BOD test and the ammonia released, which together with amonia already present, can be oxidized to NO_2^- and NO_3^- with significant oxygen consumption.

The degree to which ammonia is oxidized in the standard 5-day BOD test is largely dependent on the seeding organisms. If the seed contains significant numbers of nitrifying bacteria ammonia oxidation will occur at once. However if the seed does not contain large populations of these microorganisms then nitrification may not occur during the 5-day period of the test because the nitrifiers grow relatively slowly and may not have time to develop a significant population in 5 days.

Measurement of BOD

The BOD test is based on determinations of dissolved oxygen; consequently, the accuracy of the results is influenced greatly by the care given to its measurement. BOD may be measured directly in a few samples, but in general, a dilution procedure is required.

Two methods are used widely for BOD measurement: (1) the dilution method, which is a standard method of the American Public Health Association (APHA) and approved by the U.S. Environmental Protection Agency (EPA), and (2) the manometric method. The EPA denied approval of this method when it selected methods for wastewater analysis, although in certain cases the EPA has approved the manometric method (Hach, 1984).

The dilution method is carried out by placing various incremental portions of the sample being analyzed in bottles and filling the bottles with what is called "dilution water". The dilution water contains a known amount of dissolved oxygen. A portion of inorganic nutrients and a pH buffer are contained in the dilution water and are also provided to the samples. The bottles are filled completely full, free of air bubbles, sealed and allowed to stand for five days at a controlled temperature of 20°C (68°F) in the dark. During this period, bacteria oxidize the organic matter, using the dissolved oxygen present in the water. At the end of the 5-day period, an analysis for the remaining dissolved oxygen is made. The amount of oxygen that was consumed during the 5 days in relation to the volume of the sample increment is then used to calculate the BOD.

Measurement of BOD by the manometric method is easier because the oxygen consumed is measured directly rather than with chemical analysis. Because the sample is usually tested in its original state (no dilution), its behaviour more closely parallels that of the waste in an actual sewage-treatment plant. As the oxygen in the sample is used up, more will dissolve into it from the air space over it. The manometer, which is essentially a column of mercury, measures the drop in air pressure in the bottle. This continuous indication of the amount of oxygen uptake by the sample is a very important feature of the manometric method. By graphing the results, one can find the rate of oxygen uptake at any time and thereby gain considerable insight into the nature of the sample.

Review

KEYWORDS

Biochemical Oxygen Demand (BOD)
bioassay
seed
oxidation reaction
manometric measure

Suggested Readings

American Public Health Association, 1985. "Standard Methods for the Examination of Water and Wastewater." Published by the Amer. Public Assoc., Amer. Water Works Assoc., and Water Poll. Control Fed., 16th Ed. pp. 525–532.
Hach Chemical company, 1984. "Hach Water Analysis Handbook." Ames, Iowa.
Sawyer, C.N., and P.L. McCarty, 1978. "Chemistry for environmental engineering." 3rd Ed. New York.

Chapter 18

Enteric Bacteria

Purpose: To learn the membrane filter technique for determination of three common microorganisms: total coliform, fecal coliform, and fecal streptococcus.

The sanitary quality of our nation's waters must be protected to provide for unrestricted use of streams for domestic, commercial, and recreational purposes. Water, if not adequately protected, may not be usable as a public water supply, for fish and wildlife harvesting, or for water sports.

Sanitary quality refers specifically to the presence or absence of pathogenic (disease causing) bacteria. The fewer the number of pathogens, the higher a water's sanitary quality.

Traditionally, use has been made of tests for the detection and enumeration of indicator bacteria (bacteria types that are frequently associated with pathogens and may indicate their presence) rather than for the pathogens themselves. The principal indicator used has been the coliform bacterial group. The number of coliform bacteria in a water sample has been a good measure of sanitary quality, except that the coliform group is comprised of types that have their origin in soils and others that have their origin in fecal matter. Having these two sources causes the problem that in some instances waters may appear much poorer than they really are, in that the coliform count may be very high yet very few pathogens exist.

The established standards recognize the problems associated with using coliform bacteria as an indicator of sanitary quality. Since the coliform are indicator organisms and not pathogens, the criteria are written in terms of median values (see Chapter 20), not maximum values. Also, the criteria require that fecal sources be present before that data is used to indicate sanitary quality.

Bacterial densities in surface waters range from less than 50 per 100 mL in wildland waters to more than 10,000 per 100 mL in streams near urban and agricultural areas. Many streams seem of marginal quality, particularly when compared with water-quality criteria values. However, the criteria require that the origin and concentration must both be considered in judging true sanitary quality. Many remote wildland streams undoubtedly have low densities, but since they are frequently inaccessible and in most cases there is no reason to believe they have water quality problems, little data have been collected.

The highest coliform densities are often found near agricultural activities. Additional study is needed to determine the true sanitary and health significance of these high numbers of coliform bacteria. In the Colorado River, for example, coliform counts increase substantially in the lower river, where irrigation return flow is dominant. A large portion of these organisms may be soil rather than fecal bacteria.

Perhaps the most critical requirement for drinking-water quality is the absence of pathogens. Since it is not practical to isolate each pathogen, tests are done for groups of organisms that are indicators of the type, proximity, and degree of fecal pollution.

Total coliforms include over 30 species of differing characteristics and habitats. Nonfecal coliforms, while unacceptable in drinking water, may come from nearby or distant fecal pollution or soil runoff, and include many innocuous species.

Fecal coliforms specifically indicate recent fecal waste contamination, since this is their sole source and their survival in water is limited. The number of these organisms is greater in humans than in other warm-blooded animals and their presence in a drinking-water supply is cause for concern. The test for these organisms relies on their resistance to heat shock, which inhibits most nonfecals.

Fecal streptococcus tests are used in conjunction with the above tests to provide information on the nature of the sources of pollution.

There are generally more fecal streptococcus in warm-blooded animals other than humans.

Procedure

Sample storage and transfer time should be minimized with all bacteriological samples—maximum 6 hours, held below 10°C. One of the advantages of the membrane filter technique is that filtering can be done in the field to meet this criterion, and the filters can be shipped in special transport media that keeps the organisms viable while inhibiting growth.

The membrane filter technique is described as follows. A sterile absorbent pad 47 mm in diameter is placed in a sterile glass or plastic petri dish 50–60 mm in diameter, and about 15 mm in height. Plastic petri dishes with tightly fitting covers should be used to prevent moisture loss during incubation. The absorbent pad is saturated with a bacterial broth growth medium (e.g., 1.8–2.2 mL). The excess medium is drained off. A solid medium can also be used. In this case, no absorbent pad is needed; the liquified medium can be poured in advance directly into the petri dish, allowed to solidify, and stored under refrigeration until required.

Depending on the nature of the sample and the tests to be run, a volume of the sample must be selected to yield a number of colonies on the filter that can be enumerated with accuracy. It is a good idea to run a number of different volumes simultaneously. Large sample volumes may be divided among more than one filter; however, 30–100 mL samples for wildland water are usually sufficient.

The water sample is passed through a membrane filter (preferably grid-marked) composed of cellulose esters with a fixed pore diameter size. The filter is then placed grid side up on top of the previously broth-saturated pad or the solid medium. All the bacteria in the sample are retained on the membrane into visible colonies consisting of many individual cells. If the inoculant sample contains the proper range in bacterial density, countable plates will be evident after an incubation period. The constituents of the growth medium and the temperature of incubation will influence the type of bacterium that will develop (see Table 18.1).

USGS - autoclave

TABLE 18.1
Specific Conditions for Membrane Filter Tests

Test[a]	Medium	Incubation Temperature (°C)	Time (hours)	Colony Color
TC	MF–Endo broth	35 ± 0.5	22–24	Green or gold metallic sheen
FC	MF–FC broth	44.5 ± 0.2	24	Light blue to dark blue green
FS	KF agar	35 ± 0.5	48	Pink to red

[a] TC = total coliforms; FC = fecal coliforms; FS = fecal streptococcus.

PROCEDURE

Detailed descriptions of the various procedures are given in the Millipore Application Manual and their various brochures. Only an outline of the procedure will be given here. You should do at least three replications for each sample.

Apparatus

1. Sample bottles
2. Dilution bottles
3. Pipettes
4. Graduated cylinders
5. Culture media containers
6. Culture dishes
7. Filtration units
8. Filter membranes
9. Absorbent pads
10. Forceps
11. Incubators
12. Stereomicroscope with light source

Steps

1. Place a sterile absorbent pad in a 47 mm petri dish using smooth-tipped sterilized forceps or a millipore dispenser.
2. Add 2 mL medium through a sterile pipette.
3. Close the petri dish and set aside until step 10.

4. Sterilize a Hydrosol Filter Holder in an autoclave.
5. Load the Hydrosol holder with a membrane filter, grid side up. When handling the filter, be sure to use smooth-tipped forceps that have been dipped in alcohol, then quickly flamed to sterilize the tips, and allowed to cool a few seconds before handling the filter.
6. Add about 20 mL of sterile buffer to the filter holder funnel. The exact volume is not critical. Its purpose is only to evenly disperse the bacteria in the measured sample.
7. Into the filter holder funnel, pipette a predetermined aliquot of the sample water. Swirl the funnel to mix the sample with the sterile buffer. The size of the aliquot will vary with the contamination level of the water being sampled. An ideal sample volume will yield a certain range of colonies on the filter surface.
8. Using a vacuum pump, or some other suitable vacuum source, apply vacuum to the receiver flask. This will cause water to flow through the filter, leaving the bacteria trapped on the filter surface.
9. With the filter still in place, rinse the walls of the funnel with sterile buffer. Use a volume at least equal to the total volume of the liquid just filtered. Apply vacuum and draw rinse buffer through. Repeat this rinse.
10. After all the rinse buffer has been drawn through the filter, release the vacuum once again, remove the filter holder funnel and using flame sterilized forceps, lift the filter from the filter holder base placing it grid side up on the saturated absorbent pad in the petri dish prepared in step 3. Carefully line up the filter with one edge of the petri dish and set it down (evenly centered) with a slight rolling motion, avoiding air entrapment. Replace the cover tightly.
11. Rinse filter apparatus thoroughly with sterilized buffer solution before filtering the next sample.
12. Incubate the petri dishes upside down as specified for each test. This should be done within 30 minutes of filtration.
13. At the end of the incubation period, remove the petri dish and count the colonies, using a stereomicroscope at 10–20 magnification. The light source should be daylight fluorescent or natural daylight, and should fall nearly vertically on the culture.

CALCULATIONS

The accepted way of expressing bacteria levels in water is in terms of the number per 100 mL. Bear in mind that each colony observed developed from a single bacteria cell in the original sample. To determine the number (N) of bacteria in the sample tested, use the following formula:

$$\frac{(N \text{ counted}) \ (100)}{mL \text{ sample}} = N \ 100 \ mL^{-1} \text{ sample}$$

INTERPRETATION

Total Coliform

Coliform types will exhibit a green or gold metallic sheen in the center of the colony or around the periphery. Noncoliforms will appear light or dark red; if they are wet, a false sheen may be seen if the light source is too strong.

Fecal Coliforms

Fecal coliforms (FC) will exhibit a light blue to dark blue–green color. Nonfecals usually appear gray to cream-colored. Subtle color changes may occur outside the incubator, so observe within 30 minutes.

Fecal Streptococcus

Fecal streptococcus (FS) colonies should appear light pink to smooth clear red with pink margins.

Fecal Coliform/Fecal Streptococcus Ratio

The concentration of FC per 100 mL divided by the concentration of FS per 100 mL provides the FC/FS ratio. Generally, a ratio under 0.7 indicates animal fecal contamination; over 4.0, human fecal contamination. Ratios between 0.7 and 0.4 are usually from mixed sources. Do not apply the ratio if the fecal strep count is below 100 per 100 mL.

For the results to be valid fecal strep density should be greater than 25 100 mL^{-1} and the pH must fall between 4.0 and 9.0. Samples should be taken from the same site at the same time, as close to the source as possible. They should never be taken more than 24 hours downstream from the source.

Bacteria in streams typically have large variations in population numbers. Sampling schemes must account for this variation; otherwise conclusions drawn from a small size may produce erroneous results (Kunkle, 1970). Bacterial populations may change during the day because of varying ultraviolet light intensity; the light may kill bacteria. Time of sampling should be considered in the water-quality monitoring program. Time bias should be avoided, if travel times between sample points is long or if many sample sites are involved. Bacteria concentrations tend to increase with stream discharge; thus sampling frequencies should be planned with this flow-driven relationship considered.

TYPICAL VALUES

	Fecal Coliform	Fecal Streptococcus
	Colonies 100 mL^{-1}	
Cache la Poudre River at canyon mouth near Ft. Collins, Colorado	46	69
Tualatin River at West Linn, Oregon	183	372
Rogue River near Agness, Oregon	56	376
Colorado River at Stateline, Colorado	470	510
Colorado River at Imperial, California	51	180
Skagit River at Marblemount, Washington	69	218
Yellowstone River near Livingstone, Montana	—	70
Arkansas River near Coolidge, Kansas	310	1100
Cumberland River near Grand River, Kentucky	811	618
Ogeechee River near Eden, Georgia	51	69
North Fork Whitewater River near Elba, Minnesota	31	60

Review

KEYWORDS

coliform
streptococcus
fecal bacteria
membrane filter
incubation
enteric bacteria
microbial diseases

STUDY QUESTIONS

1. In some areas remote from human habitation, there may be large concentrations of coliform bacteria in streams. How would you explain this?
2. Do colonies of bacteria growing on a nutrient medium give a quantitative estimate of the size of natural microbial populations? Why or why not?
3. Calculate the average bacterial density in your samples, for total coliforms, fecal coliforms, and fecal streptococcus.
4. Discuss the potential effects of discharging sewage effluents onto land, rather than into water courses. Describe the effects of soil, climate, and topography on the fate of bacteria in such a land application procedure.
5. What effect might multiple point discharges into a stream, such as sewage effluents and heated water from a power plant, have on the quality of water in the stream?

SAMPLE DATA SHEET

Analysis: Enteric Bacteria   Instrument name:
Analyst: Instrument serial number:
Date: Calculations done by:
Start time: Calculations checked by:
  End: Comments:
Procedure:

SAMPLE IDENTIFICATION:	REPLICATE	MEASUREMENT
Standard	1	
	2	
Reported Value:	3	
	Mean:	
	Standard Deviation:	
	In Control?	

SAMPLE IDENTIFICATION:	REPLICATE	MEASUREMENT
	1	
	2	
Reported Value:	3	
	Mean:	
	Standard Deviation:	
	In Control?	

SAMPLE IDENTIFICATION:	REPLICATE	MEASUREMENT
	1	
	2	
Reported Value:	3	
	Mean:	
	Standard Deviation:	
	In Control?	

Suggested Readings

American Public Health Association, 1985. "Standard Methods for the Examination of Water and Wastewater." Published by the Amer. Public Health Assoc., Water Works Assoc., and Water Poll. Control Fed., 16th Ed.

Aukerman, R., and W.T. Springer, 1976a. *Effects of recreation on water quality in windlands.* Eisenhower Consortium Bull. 2, USDA Forest Service, Rocky Mountain Forest and Range Experiment Station, Fort Collins, Colorado. 25 pp.

Aukerman, R., and W.T. Springer, 1976b. *Effects of recreation on water quality in wildlands.* Eisenhower Consortium Bulletin No. 2. NTIS. Springfield, Virginia 25 pp.

Barton, M., 1969. Water pollution in remote recreation areas. *J. Soil and Water Conserv.* **24**: 132–134.

Bordner, R., J. Winters, and P. Scarpino (eds.), 1978. *Microbiological methods for monitoring the environment: Water and wastes.* EPA-600/8–78–017. U.S. Environmental Protection Agency, Environmental Monitoring and Support Laboratory, Cincinnati, Ohio. 337 pp.

Brown, G.W., 1983. "Forestry and water quality." (2nd Ed.) Oregon State University Bookstores, Inc., Corvallis, Oregon. 142 pp.

Burton, G.A, 1982. *Microbiological water quality of impoundments: A literature review.* Miscellaneous Paper E-82–6. U.S. Army Corps of Engineers, Waterways Experiment Station, Vicksburg, Mississippi. 53 pp.

Cabelli, V.J., 1983. *Health effects criteria for marine recreational water.* EPA-600/1–80–031. U.S. Environmental Protection Agency, Health Effects Research Laboratory, Research Triangle Park, North Carolina. 98 pp.

Geldreich, E.E., 1970. Applying bacteriological parameters to recreational water quality. *J. Amer. Water Works Assoc.* **62**: 113–120.

Kabler, P.W. and H.F. Clarke, 1960. Coliform group and fecal coliform group organisms as indicators of pollution in drinking water. *J. Amer. Water Works Assoc.* **52**: 1577–79.

Kunkle, S.H., 1970. Concentrations and cycles of bacterial indicators in farm surface runoff. *In* "Relationship of Agriculture to Soil and Water Pollution. Proceedings, Cornell University Conference on Agricultural Waste Management," Ithaca, New York, Jan. 19–21, 1970. pp. 49–60.

Kunkle, S.H., 1972. Sources and transport of bacterial indicators in rural streams. *In* "Proceedings, Interdisciplinary Aspects of Watershed Management," Bozeman, Montana, Aug. 3–6, 1970. American Society of Civil Engineers, New York.

Millipore Corporation, 1975a. AB314 Field Procedures in Water Microbiology. Bedford, Massachusetts.

Millipore Corporation, 1975b. AB315 Fecal Streptococcus Analysis. Bedford Massachusetts.

Millipore Corporation, 1975c. AM302 Biological Analysis of Water and Wastewater. Bedford, Massachusetts.

Millipore Corporation, 1974. AB313 Fecal Coliform Analysis. Bedford, Massachusetts.

Tunnicliff, B., and S.K. Brickler, 1984. Recreational water quality analyses of the Colorado River corridor in Grand Canyon. *Applied and Environmental Microbiology* **48**: 909–917.

Chapter 19

Contract Laboratory Selection

Laboratory Selection

The analytical techniques presented in this text are for the more common constituents associated with land-use and water quality studies. The wildland hydrologist may not have access (or need) for such laboratory equipment and may elect to contract analytical services.

Recent legislation regarding leaking underground storage tanks (LUSTs), hazardous wastes, and other site remediation requirements has resulted in proliferation of analytical laboratories. The choice of contract lab should be made after evaluation of the quality-assurance program, certification, sample turnaround time, and costs.

Quality-Assurance Program

Quality-control programs ensure a specified degree of confidence in the analytical results. Various federal guidelines have been developed for quality-assurance programs and include the following considerations.

Routine analysis of a standard sample along with analyses of unkown samples to check the accuracy of the results. Standard

Figure 19.1 Paired water-quality samples (time and space) may be submitted and analyzed as split samples. Quality-control measures should be included for both field and lab analyses.

samples may be run from one every three samples to one every 20 or more samples. If the standard sample result is out of control (usually within 5 percent of the mean), the unkown samples are reanalyzed.

Reference samples are obtainable from the U.S. Environmental Protection Agency (EPA) or the National Bureau of Standards. Reference samples are used to evaluate analytical techniques.

Large labs often have a quality control officer who will run split samples (see Figure 19.1) and/or spike samples as well as the standard and reference samples. You should be informed of the quality-assurance program and the error limits for out of control samples.

CERTIFICATION

If a laboratory is EPA- or state-certified, quality control is maintained to the certification standards. Certified laboratory results are usually more expensive than noncertified laboratory results. However, noncertified laboratories are capable of producing results just as accurate. If there is the possibility that the results will be used in any legal proceedings, it is easier to have results entered as evidence if the laboratory is certified.

SAMPLE TURN AROUND TIME

Sample turnaround time should be part of your consideration for lab selection. If sample results are needed for future monitoring decisions, or if the results are for inventory or background information only, the sample submission, especially via mail, should be evaluated.

Sample turnaround times of more than 2 months are becoming common for some labs. Samples 2 months old or older would pose a logistical problem for storage as well as provide opportunities for sample degradation or constituent transformation. If laboratories are changed during the course of the study, be certain the same analytical techniques and reporting accuracy are used. If one lab uses two significant figures and the other lab uses four significant figures, any final reporting of results must use the more conservative figures.

COSTS

You do not necessarily get what you pay for. The range in analytical costs is wide, and the user should beware. Standard analytical techniques may be offered in a routine package. Also, volume (sample number) discounts are available. Standard or routine analytical packages may be cheaper and provide more information than selecting only a few parameters of interest.

The price list should include detection limits. Parameters of interest should not be below the detection limit if meaningful statistics are to be employed. Conversely, if milligrams per liter is sufficient, results need not be reported to micrograms per liter.

Chapter 20

Data Analysis and Presentation

Statistical Methods

Statistics is the study of methods and procedures for collecting, classifying, summarizing, and analyzing data, and for making scientific inferences from such data. There are two basic types of statistics: descriptive and inferential.

DESCRIPTIVE STATISTICS

The use of graphs and tables serve as devices for organizing data and representing essential characteristics for the purpose of reaching conclusions at a later stage. Parametric and nonparametric statistics may be used to characterize the results.

INFERENTIAL STATISTICS

A basic characteristic of experimental surveys is the necessity for reaching conclusions regarding a population on the basis of studying only a sample of that population. The population is the full set of elements to which we limit any discussion or inference, while the sample is a subset of that population. Inferential statements are usually limited to the quantitative aspects of generalization, although often they contribute to the process of reaching conclusions.

ACCURACY, PRECISION, AND BIAS

In statistics, "accuracy" refers to the success of estimating the true value of a quantity; "precision" refers to the spread of sample values about their own mean. An example for water-quality professionals might concern replicated measurements of stream water pH at a given location. If the measurements are carefully made, they should not vary a great deal but should cluster closely about their mean value; they will be precise. However, if the pH meter is not properly calibrated the measured values will be inaccurate. If the pH meter is properly calibrated and is used carefully, the measurements will be both accurate and precise.

"Bias" is a systematic distortion caused by a flaw in measurement, method or sample selection, or techniques used in estimating a parameter. An example of a bias would be the selection of a stream sampling site based solely on its easy access. Obviously, some biases may introduce more error than others.

SAMPLE PARAMETERS

The objective of a sample survey is to estimate some characteristics of a population without measuring all the population elements. Thus, the two basic objectives of statistical methods are (1) estimation of population parameters (any of a set of physical properties whose values describe the characteristics of a particular population) and (2) hypothesis testing about these parameter estimates.

A random sample is one that when all elements of size n are selected that every possible combination of n elements has an equal chance of being selected. Oftentimes conditions are such that measurements or samples are taken arbitrarily and not randomly as required by most statistical techniques. All one can hope for is that those arbitrary measurements or samples act as well as random ones.

POPULATION PARAMETER ESTIMATES

Mean

For a given random sample, the sample mean is the best estimator of the unknown population mean. The mean $\bar{x}$ is the sum of sample values divided by the number of values in a sample of size n. It may be expressed as

$$\bar{x} = \frac{\Sigma x_i}{n}$$

where Σx_i is the sum of observed values of the i_n sample and n is the number of values in the sample.

If the analysis has values reported below detection, the user may elect to use 50 percent of the detection limit as the numeric value. This has a 50 percent chance of being too high and 50 percent chance of being too low. If the population is log-normally distributed (e.g., heavy metals), use 70 percent of the detection limit.

Median

The median value of a sample may be more appropriate than the mean. This is particularly true when the sample appears to have a log-normal distribution but is too small to be tested, or when several analyses are below the detection limit. If the sample size is odd, the median is the middle term when the samples are arranged in increasing order. When the sample size is even, the median is the average of the two middle terms.

Standard Deviation

The standard deviation characterizes variation between individuals and gives an idea as to how elements are distributed about the mean. For a random sample of n units, the estimated population standard deviation is:

$$s = \sqrt{\frac{\Sigma(x_i - \bar{x})^2}{n - 1}}$$

where x is the sample mean and $(x_i - \bar{x})$ indicates the difference between individual measurements from the mean.

Coefficient of Variation

In natural systems, populations with large means tend to show more variation than do populations with small means. The coefficient of variation (CV) enables you to compare variability about these different-sized sample means. It is estimated by dividing the standard deviation by the mean. The coefficient of variation is usually expressed as a percent:

$$CV = \frac{s}{\bar{x}} \times 100$$

Standard Error of the Mean

The standard deviation is a measure of the variation among the individual units of a population. Variation may also exist among the means of different samples taken from a single population. The standard error of the mean ($S\bar{x}$) is an estimate of the variation likely to be encountered among the means of different samples from a population. It can be thought of as the variation among sample means just as the standard deviation is the variation among elements of a sample. If the sample size is very small as compared to the population size (less than 5 percent), the standard error of the mean is estimated by

$$S\bar{x} = \sqrt{\frac{s^2}{n}}$$

where s^2 is the variance, which is equal to the square of the standard deviation, and n is sample size.

Confidence Limits

Estimated sample means of a population may vary. The degree of variation depends primarily on the inherent population variability and the sample size. A method of indicating the reliability of a sample mean is to establish confidence limits. Although confidence limits can be determined for any sample estimate, only confidence limits on the mean will be covered. For normally distributed populations, the confidence limits are expressed as

$$\bar{x} \pm (t)\,(S\bar{x})$$

where $\bar{x}$ = sample mean, t = value of Student's t obtained from the t-distribution table (see Table 20.1; r = degree of freedom = $n-1$),

TABLE 20.1
Student's t Critical Values[a]

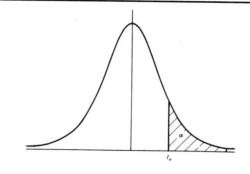

			α			
r	.25	.10	.05	.025	.01	.005
1	1.000	3.08	6.31	12.7	31.8	63.7
2	.816	1.89	2.92	4.30	6.97	9.92
3	.765	1.64	2.35	3.18	4.54	5.84
4	.741	1.53	2.13	2.78	3.75	4.60
5	.727	1.48	2.02	2.57	3.37	4.03
6	.718	1.44	1.94	2.45	3.14	3.71
7	.711	1.42	1.89	2.36	3.00	3.50
8	.706	1.40	1.86	2.31	2.90	3.36
9	.703	1.38	1.83	2.26	2.82	3.25
10	.700	1.37	1.81	2.23	2.76	3.17
11	.697	1.36	1.80	2.20	2.72	3.11
12	.695	1.36	1.78	2.18	2.68	3.05
13	.694	1.35	1.77	2.16	2.65	3.01
14	.692	1.35	1.76	2.14	2.62	2.98
15	.691	1.34	1.75	2.13	2.60	2.95
16	.690	1.34	1.75	2.12	2.58	2.92
17	.689	1.33	1.74	2.11	2.57	2.90
18	.688	1.33	1.73	2.10	2.55	2.88
19	.688	1.33	1.73	2.09	2.54	2.86
20	.687	1.33	1.72	2.09	2.53	2.85
21	.686	1.32	1.72	2.08	2.52	2.83
22	.686	1.32	1.72	2.07	2.51	2.82
23	.685	1.32	1.71	2.07	2.50	2.81
24	.685	1.32	1.71	2.06	2.49	2.80
25	.684	1.32	1.71	2.06	2.49	2.79
26	.684	1.32	1.71	2.06	2.48	2.78
27	.684	1.31	1.70	2.05	2.47	2.77
28	.683	1.31	1.70	2.05	2.47	2.76
29	.683	1.31	1.70	2.05	2.46	2.76
z	.674	1.28	1.645	1.96	2.33	2.58

[a] For df $\geq$ 30, the critical value t_α is approximated by z_α, given in the bottom row of table.

and $S\overline{x}$ = standard error of the mean. For the probability column of the t-table, choose a value of 0.05 to 0.01, for 95 or 99 percent probability, respectively. The probability value indicates the range of the confidence limit estimate. Confidence limits (or size) increase with increased probabilities.

Sample Size

To optimize a water quality survey, a sufficient number of samples should be taken to obtain the desired precision—no more, no less. Although samples cost money, an insufficient number of samples to conduct reliable statistical analyses based on your objectives may cost even more as well as contributing to a waste of time and energy.

The more variable a water quality constituent is in time and space, the more frequently it must be sampled to achieve a given level of reliability. The following method enables one to estimate a minimum sample size for water quality sampling, when a comparison of means is required and when the mean is estimated from a series of random samples.

An iterative method used in solving for a desired sample size (n) is

$$n = \frac{t^2 \, s^2}{E^2}$$

where t = the value from the Student's t-table at a certain level of probability, s = the standard deviation, and E = the specified error.

Example

The objective of the iterative process described here is to obtain a predicted n approximately equal to the calculated n. For $t_{0.025}$, $s = 1.414$, and $E = 2$ mg L^{-1} of D.O. for example.

At $n = 10$ (predicted), $t_{0.025,9} = 2.26$ and $(2.26)^2(1.414)^2/2^2 = 2.55$. Try a lower n.

At $n = 5$ (predicted), $t_{0.025,5} = 2.78$ and $(2.78)^2(1.414)^2/2^2 = 3.86$. This is closer, but try lower.

At $n = 4$ (predicted), $t_{0.025,4} = 3.182$ and $(3.18)^2(1.414)^2/2^2 = 5.06$. Since $4 < 5.06$, this is close enough.

From this iterative procedure for determining a sample size, you found that the sample size should be between four and five samples to achieve a specified error E of 2 mg L^{-1} DO. Always be conservative and take the higher value, in this case, five samples.

If the preliminary sampling has been completed, we can estimate s^2 from the data. Otherwise, an approximation of the sample variance may be calculated from the following:

$$s^2 = \left(\frac{R}{4}\right)^2$$

where s^2 is variance and R is range (maximum–minimum concentrations expected). The range may be determined by professional judgment or by preliminary sampling data.

Data Distributions

A distribution function shows the relative frequency with which different sampled values of a parameter occur. For each parameter of a given population there is a unique distribution function. The bell-shaped normal distribution is one type of function that frequently approximates most population distribution functions. The statistical methods developed around the normal distribution are simpler than those developed for other distributions. Despite the distinct distributions each parameter follows, the means of large samples tend to approximate a normal distribution and may be treated by normal distribution statistical methods.

This section will address procedures for determining whether a sample data set is normally distributed, and common methods for changing or transforming nonnormal data sets into normal distributions.

PROBABILITY PAPER PLOTS

One of the easiest methods of determining whether a data set is normally distributed is to plot the data values on probability paper. A normally distributed data set will plot as a straight line. A problem arises when you must decide just how far removed your data are from a straight-line plot before your data set is no longer acceptable to be considered normally distributed. Two quantitative measures of data-set normality are skewness and kurtosis.

SKEWNESS

The normal distribution is symmetric about its mean value. A distribution that is asymmetric (not symmetric) can be skewed to the left (negatively skewed) or skewed to the right (positively skewed). The skewness coefficient (G) can be calculated as

$$G = \frac{n\Sigma(x_i - \bar{x})^3}{(n-1)(n-2)s^3}$$

where n = sample size
  $\Sigma(x_i - \bar{x})^3$ = sum of the cubed individual data-set values minus the sample mean
  s = the standard deviation

If $-1.0 \leqslant G \leqslant 0.5$, then the sample value is normally distributed; if $G > 0.5$, right-skewed; and if $G < -1.0$, left-skewed.

KURTOSIS

"Kurtosis" describes the degree of peakedness of a data-set distribution in relation to the length and size of its tails. The kurtosis coefficient (K) can be calculated as

$$K = \frac{n^2 \Sigma(x_i - \bar{x})^4}{(n-1)(n-2)(n-3)s^4}$$

where n = sample size
  $E(x_i - \bar{x})^3$ = sum of the cubed individual data-set values minus the sample mean
  s = the standard deviation

if $K = 2 - 4$, the distribution is normally peaked (mesokurtic); $K < 2$, it is flat (platykurtic); $K > 4$, it is highly peaked (leptokurtic).

TRANSFORMATIONS OF SAMPLE DATA

Data sets that do not follow something close to a normal distribution function can often be changed or transformed by various methods to achieve normality. Three methods of transformation commonly used in water quality statistics are $\log x$, $x + 1$, $\sqrt{x}$, and x^2 (where x = a sample value). Tests for normality on transformed data can then be performed to determine whether the new distribution function is normal. Data sets that do not approximate a normal distribution even after a transformation can still be analyzed using nonnormal statistical methods. These methods can be found in most statistics texts.

Hypothesis Testing

The role of probability is to enable you to put confidence limits on an estimated parameter. A common misconception is that "You can prove anything with statistics." Actually, you cannot prove anything. At best you can estimate the probability of something happening and as a scientist draw conclusions based on statistical analyses.

Various levels of reliability can be selected for probability testing. If you always test at the 0.05 level [alpha (α) level: = 0.05], you will make a mistake on the average of 1 time in 20 (95 percent reliability) of rejecting a hypothesis that is true or accepting one that is false. Should you wish to lower your degree of risk, a level of 0.01 (99 percent reliability), may be chosen. Although the choice is yours, a level of 0.05 is commonly selected in water quality sampling statistics.

STUDENT'S t-TEST

The t-test is a statistical method for testing the hypothesis that there is no difference between treatment (or group) means. This section will address comparisons between two groups by the t-test for paired and unpaired samples.

Paired Samples

The paired-samples test involves measurements that were taken exclusively in pairs. The measurements are obtained by restricted random sampling. The testable hypothesis is that there is no real difference between the two group means ($H_0: \bar{x}_A = \bar{x}_B$). The calculated value of (t) when the groups have been paired is

$$t_{calc} = \frac{\overline{X}_A - \overline{X}_B}{(s^2d)^{1/2}/n}$$

with $(n - 1)$ degrees of freedom

where $\overline{X}_A, \overline{X}_B$ = the mean of groups A and B
n = sample size
s^2d = the variance of the paired-sample differences (d)
between groups A and B

$$s^2d = \frac{\Sigma d^2 - \dfrac{(\Sigma d_i)^2}{n}}{n - 1}$$

The testable (null) hypothesis is H_0: mean A = mean B. If the testable hypothesis is false, then the alternate hypothesis is accepted H_a: mean A $\neq$ mean B. For a selected α value, the tabular (t) α value is $\alpha/2$ (two-tailed).

If the calculated t value is greater than the tabular (t), value then there is a significant difference between group means at a given α level.

Example

A test is conducted to determine whether dissolved-oxygen concentrations are the same above and below a small reservoir. Eleven paired water samples are taken from the stream above and below the reservoir with the following results:

	1	2	3	4	5	6	7	8	9	10	11	sum	mean
Above	12	8	8	11	10	9	11	11	13	10	7	110	10
Below	10	7	8	9	11	6	10	11	10	8	9	99	9
$A_i - B_i$	2	1	0	2	−1	3	1	0	3	2	−2	11	1

For a selected alpha value of 0.05

$$s^2d = \frac{\Sigma d_i^2 - (\Sigma d_i)^2/n}{n - 1} = \frac{2^2 + 1^2 + \cdots + (-2)^2 - 11^2/11}{10} = 2.6$$

$$t_{calc,10} = \frac{\overline{X}_A - \overline{X}_B}{\sqrt{\dfrac{s^2d}{n}}} = \frac{10 - 9}{\sqrt{2.6/11}} = 2.06$$

We will accept that the dissolved oxygen means are different only if the calculated t is greater than the tabular t:

$$t_{calc,10} = 2.06 < t_{0.025,10} = 2.23$$

There is no significant difference between the means at the 0.05 level of confidence. In other words, we are 95 percent confident that there is no difference in dissolved-oxygen concentrations above and below the reservoir.

Unpaired Samples

The unpaired-samples test involves measurements that are taken as completely random sampling. The testable hypothesis is the same as for paired sampling in that you are trying to ascertain whether a significant difference between the two group means occurs. The calculated value of (t) for unpaired groups is

$$t = \frac{\overline{x}_A - \overline{x}_B}{\sqrt{\dfrac{s_p^2(n_A + n_B)}{n_A n_B}}}$$

where $\overline{x}_A$ and $\overline{x}_B$ = the means for groups A and B, n_A and n_B = the number of elements in groups A and B, and s^2 = the pooled within-group variation (calculation shown in example problem).

Example

Determine whether there is any difference in the dissolved-oxygen concentration in streams A and B given the following sample measurements:

	1	2	3	4	5	6	7	8	9	10	11	sum	mean
Stream A	11	8	10	8		10	8	8	9	11	11	94	9.4
Stream B	9	8	6	10	6	13		7				59	8.4

For a selected α value of 0.05, first calculate the corrected sum of squares (SS) within each group:

$$SS_A = \Sigma x_A^2 - \frac{(\Sigma x_A)^2}{n_A} = 11^2 + 8^2 + \cdots = 11^2 - \frac{(94)^2}{10} = 16.4$$

$$SS_B = \Sigma x_B^2 = \frac{(\Sigma x_B)^2}{n_B} = 9^2 + 8^2 + \cdots + 7^2 - \frac{(59)^2}{7} = 37.7$$

Then calculate s_p^2:

$$s_p^2 = \frac{SS_A + SS_B}{(n_A - 1) + (n_B - 1)} = \frac{16.4 + 37.7}{(10 - 1) + (7 - 1)} = 3.6$$

Finally, determine the calculated t value:

$$t_{calc} = \frac{\overline{X}_A - \overline{X}_B}{\sqrt{\dfrac{s_p^2(n_A + n_B)}{n_A n_B}}} = \frac{9.4 - 8.4}{\sqrt{\dfrac{3.6(10 + 7)}{10(7)}}} = 1.07$$

The tabular t value has $(n_A - 1) + (n_B - 1)$ degrees of freedom. If it exceeds the calculated t value at a selected probability level, we would accept the hypothesis that the means are not significantly different.

$$t_{calc} = 1.07 < t_{0.025, 17} = 2.11$$

Since the calculated t value is less than the tabular t value, the difference is not significant at the 0.05 level. Therefore the dissolved-oxygen concentrations do not differ.

Decision Making

Sampling enables you to quantitatively describe water-quality parameters and make predictive statements regarding these parameters based on statistical analyses.

Water-quality sampling is often more systematic than random. The majority of water-quality sampling surveys have been conducted by some form of systematic sampling at selected locations. There are two reasons that may justify these biases: (1) sampling deliberately spread over a selected interval of time is a good representation of the temporal variability of a natural system, and (2) sample site locations are best determined considering potential sources of disturbances, logistical support, and accessibility.

Previous use of normal distribution statistics has given water quality scientists reliable estimates despite the known hazards of nonrandom sampling methods.

Testing of hypotheses with statistics enables the wildland hydrologist to accept or reject the hypotheses with a certain confidence. The decision or conclusion that is made must recognize that there is always a certain degree of uncertainty. The statistical design of any water quality monitoring must be done as part of the monitoring plan design. It is difficult to make any decision if statistics are done after the data have been collected.

Report Format

The person who writes the completion report, or article or other final product should have an active role in the fieldwork. It is sad how often reports are written after the study by someone else who only has the data and no first-hand experience of the study.

Remember the audience and the purpose of the report. It may be necessary to write more than one report. Time is rarely available to allow the writer to do the best job, but if the audience and purpose are knowingly included in the writing, the overall utility will be increased.

The use of graphics (see Figures 20.1 and 20.2 for examples), statistical summaries (Table 20.2), or other computer-generated graphics or tables requires the inclusion of complete titles. The

TABLE 20.2
Lower Hourglass Creek dissolved oxygen measurements

Sample Number	DO (mg L^{-1})
1	9
2	9
3	11
4	9
5	7
6	7
7	10
8	8
9	9
10	11

Data set:
Mean = 9
Standard deviation = 1.4
Standard error of mean = 0.4
Coefficient of variation = 15.5 percent
Range = 7 − 11
Median = 9.
Measurements by Hach High-Range DO kit.

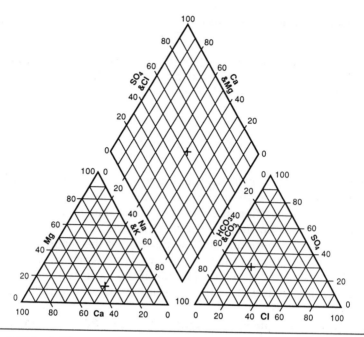

Figure 20.1 A Piper or trilinear diagram is a graphical representation of the chemistry of a water sample. Individual samples (+) may be plotted to allow for visual comparison (see Hem, 1985).

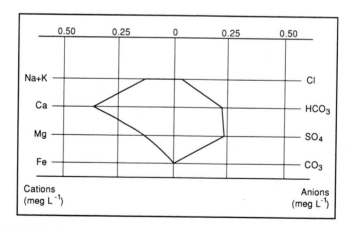

Figure 20.2 A Stiff diagram is another graphical representation of the chemistry of a particular water sample. The diagram shape and size may be used for comparisons between samples (see Hem, 1985).

graphics or tables should be able to stand on their own, that is, without supporting text; conversely, the text should be self-explanatory without the graphics. Graphics or tables should be comparable to an oral-presentation format, simple, uncluttered, and easy to read. Raw data may be included as an appendix, rarely in the main text body.

If the report or article is important enough, professional editing time should be included in the budget.

Review

KEYWORDS

descriptive statistics
inferential statistics
accuracy
precision
bias
sample parameters
mean
standard deviation
coefficient of variation
standard error of the mean
confidence limits
sample size
distribution functions
probability paper plots
skewness
kurtosis
transformations
probability testing
t-test for paired data
t-test for unpaired data

Suggested Readings

Cochran, W.G., 1977. "Sampling Techniques." 3rd Ed. John Wiley and Sons, New York. 428 pp.

Freese, F., 1967. "Elementary Statistical Methods for Foresters." U.S. Department of Agriculture, Agr. Handbook 317. 87 pp.

Freese, F., 1962. "Elementary Forest Sampling." U.S. Department of Agriculture, Agr. Handbook 232. 91 pp.

Golterman, H.D. (Ed.), 1969. "Methods for chemical analysis of Fresh Waters." International Biological Programme. Handbook No. 8. 166 pp.

Hach Chemical Company, 1981. "Hach Water Analysis Handbook." 1981 Ed. Ames, Iowa.

Hem, John D., 1985. *Study and interpretation of the chemical characteristics of natural water.* U.S. Geological Survey Water Supply Paper 2254. Washington D.C. 263 pp.

Huntsberger, D.V., and P. Billingsley, 1977. "Elements of statistical inference." 4th Ed. Allyn and Bacon, Boston. 385 pp.

Jenkins, D., V.L. Snoeyink, J.F. Ferguson, and J.O. Leckie, 1980. "Water Chemistry: Laboratory Manual." 3rd Ed. 183 pp. Wiley, New York.

Montgomery, H.A.C., and I.C. Hart, 1974. The design of sampling programmes for rivers and effluents. *Water Pollut. Control.* **66**: 77–101.

Ponce, S.L., 1980. *Water quality monitoring programs.* U.S. Department of Agriculture Service. Watershed Systems Development Group. Technical Paper. WSDG-TP-00002. 66 pp.

Remington, R.D., and M.A. Schork, 1970. "Statistics with Applications to the Biological and Health Science. "Prentice-Hall, Englewood Cliffs, New Jersey. 418 pp.

Sanders, T.G., R.C. Ward, J.C. Loftis, T.D. Steele, D.D. Adrian, and V. Yevjevich, 1983. "Design of Networks for Monitoring Water Quality." Water Resources Publications, Littleton, Colorado. 328 pp.

Sawyer, C.N., and P.L. McCarty, 1978. "Chemistry for Environmental Engineering." McGraw-Hill, New York. 532 pp.

Snedcor, G.W., and G.W. Cochran, 1980. "Statistical methods." (7th Ed.) The Iowa State University Press, Ames, Iowa. 507 pp.

Stednick, J.D., 1987. The effects of hydrochemistry variability on water quality sampling frequencies in an alpine-subalpine forested basin. *Forest Hydrology and Watershed Management.* IAHS Publ. No. **167**:401–407.

Steel, D., and J.H. Torrie, 1980. (2nd End.) "Principals and Procedures of Statistics." McGraw-Hill, New York. 663 pp.

Steel, R.G.D., and J.H. Torrie, 1960. "Principles and Procedures in Statistics." McGraw-Hill, New York. 481 pp.

United Nations Educational, Scientific and Cultural Organization (UNESCO), 1978. "Water Quality Surveys: A Guide for the Collection and Interpretation of Water Quality Data." IHD-WHO Working Group on the Quality of Water, Paris, France. 350 pp.

Index